Submarine Propulsion

Muscle Power to Nuclear

Anil Anand

Contributors:

France : Yves M. HENON,

Brazil : Fernanda das Graças Corrêa and Leonam dos Santos Guimarães

Argentina: Alejandro Delaygue

Published by

Frontier India Technology

No 22, 4th Floor, MK Joshi Building, Devi Chowk, Shastri Nagar,
Dombivli West, Maharashtra, India. 421202
http://frontierindia.org

CONTENTS

PROLOGUE AND GRATITUDE

PROLOGUE AND GRATITUDE

After the reactor, on board the first Indian indigenous nuclear submarine Arihant was made 'critical' in 2013, I was invited to give 'The Founder's Day' lecture on October 30, Dr. Bhabha's birthday which is celebrated as founder's day of BARC (Bhabha Atomic Research Center) named after Dr. Bhabha's death. I prepared and gave the lecture along with a power point presentation on the 'Evolution of Submarine Propulsion' It was highly appreciated and I was told by many of my former colleagues that I should write a simple and not highly technical book for the scientific and engineering community. I did write a book but not on this topic; I was so fascinated and was 'en amour' by my personal life & family history and my wife & her sacrifice, who had left her country & the medical profession and joined me in Mumbai to complete my mission of forty years with the Atomic Energy, that the book became my autobiography called 'The Second Strike' which was launched by then Director BARC, Dr S Basu, on September 14, 2014. And later it was released by admiral Bhushan in Delhi on November 21. A few months back, Mr. Joseph P Chacko approached me and published and printed the same book as 'Ma Dame - A Nuclear Scientists tryst with love and fission' for the overseas markets. In addition to my former innumerable colleagues, friends and well-wishers, now I have another friend Joseph P. Chacko who joined them in coaxing me to write this book; I owe my gratitude to all of them as I finish writing this book.

I thank the Chairman Atomic Energy Dr. S Basu and Director BARC Mr. K N Vyas for writing the Forewords and also agreeing to formally release this book.

I thank my former colleagues Mr R K Garg, Dr. Dwivedi, Mrs Roy, Vyas, Bhattacharya, R P Singh, A B Mukherjee and Pradip Mukherjee for their contribution. I thank Yves M Henon, Fernanda das Graças Corrêa & Leonam dosSantos Guimaraes and Alejandro Delaygue for their contribution on the Nuclear Propulsion in France, Brazil and Argentina respectively. I also thank Vinay Karanam, P Sreenivas, Tessie George and Pradip Mukherjee who put an effort to trace the references in the BARC library for me, based only on sketchy information provided by me. Most of the remaining information has been gathered from the internet, Google and Wikipedia. I salute the persons who provide this information on the net for persons like us and for the students who had to struggle hard to get the required answers to the queries. Though it took me > 100 hours to locate the precise information, but it is all worth; thus the references from the internet are not listed, if the reader wishes to go into some details he/she can go to the net.

Anil Anand, Mumbai, akanand9.11@gmail.com

FOREWORD

'The second strike' was a tale of two families; Anand Family and PRP Family. You already know about the Anand Family and now it is time to know, what PRP Family did. So, the second book. It is now a story of a journey underwater and journey down the memory lane. You all know that all forms of life evolved underwater. When you look at sea, you get many feelings. It is endless, it is wavy, it is blue. As you move along, the feelings turns into curiosity; how would it feel to float in it? How deep is it?, What is there down below? Humans had been trying to answer these questions since their existence. They dived in on their own, they used submersibles of all shapes and sizes. Quickly they realised they have to propel and they have to vary the buoyancy to move in the third dimension.

Over a couple of hundred years, these submersibles developed into the submarines of today; now she can remain underwater for long, she can propel, she can sustain human life underwater and she can fight war for her master. Like many other major innovations of the mankind, submarines of today were developed and matured during the two world wars. Development started with amateur innovation and daring explorers, and finally reached maturity in the hands of hardcore engineers and determined submariners. This completed the first two stages of submarine evolution.

You all know about India's three-stage nuclear power programme (First Stage being U235 based thermal reactors, followed by Pu239 based Fast Breeder Reactors and the third phase will have Th232 – U233 reactors). The submarines also had a third stage in their evolution. In the book, Shri Anil Anand has brought out in a lucid, factual and informative manner the third phase of submarine evolution. As you go along, you will not become an expert, but will have sufficient insight to appreciate the complexity of this technology.

This is where the second journey begins; journey down the memory lane. This programme is fast approaching its 50th anniversary, if you take the beginning of 12th batch of BARC Training School as the starting date. Imagine the number of individuals who occupied the stage during this long journey. Anand saab has made gallant efforts, and quite successfully, so, to bring these characters alive. I can see and feel all the scenes in that era being enacted in front of me.

This foreword will not be complete without mentioning about two events. After the PRP Plant was commissioned on 22nd September 2006, all the generations involved in this programme, from all organizations, gathered together. Seniors narrated the evolution process to the younger generations from their memoir. It is very difficult to explain the feelings of that day in words. I can only say it is the same feeling that parents have after their child is born. It is a feeling of being a creator.

Exactly one week after Arihant launch, PRP's existence, so far a secret was announced by inviting the media to the plant site. Next day's newspapers projected PRP as the father of Arihant. This was the recognition of indigenous capability for developing complex technology from scratch. This is a story of struggle and conflict, sorrow and joy, success and failure and in the end the pleasure of doing a job for the nation, dwarfed everything else.

This book also brings out a special capability of the author; that is to bring a set of individuals together, with very different background, to deliver a complex high technology project, of mammoth size to the country.

I wish you happy reading.

Sekhar Basu
Chairman Atomic Energy Commission &
Secretary Department of Atomic Energy

FOREWORD

It is a well acknowledged fact that nuclear submarines are among the most complex platforms ever made by engineers. This is due to the requirements for reliability, ability to withstand very high loads in an event of a conflict, ability to survive in totally marooned sea environment and the need for stealth. If a nuclear submarine is required to meet above specifications, it is essential that the nuclear power-pack also needs equally stringent design requirements. The very fact that there are only few countries, which have mastered the technology, is itself a proof that the design complexities involved are huge.

The complex design of a nuclear propulsion plant would necessarily involve contribution from a large number of persons. However, the fact that, Shri A.K. Anand has played a key role in establishing programme for nuclear propulsion in India, gives him an excellent overall perspective. The book describes very lucidly the struggle faced during the development phase. It also goes to the credit of Shri Anand that he has been able to request specialists from other countries like France, Brazil and Argentina to describe about their own development efforts which I am sure will widen the perspective of readers.

Shri Anand, whom I met for the first time in August 1979, is my first boss as Head, Fuel Design and Development and we all fondly call him Anandsaab. I personally have known Anandsaab to be a very enthusiastic and energetic person. The very fact that Ananadsaab has written this book at an age of 76 and 15 years after retirement is a proof of his enthusiasm towards any work that he takes up.

I hope that this book will prove to be a key document which fills-up important footnotes in the history of development of nuclear propulsion in India.

K N Vyas
Director Bhabha Atomic Research Centre

SUBMERSIBLES — THE HUMAN CURIOSITY

The concept of an underwater boat has roots deep in antiquity. Although there are images of men using hollow sticks to breathe underwater for hunting, the first known military use, is of divers being used to clear obstructions during the siege of Syracuse (about 413 BC), according to the History of the Peloponnesian War. At the siege of Tyre in 332 BC, divers were again used by Alexander the Great. Later legends from Alexandria, Egypt, in the 12th century AD, suggested that he had used a primitive submersible for reconnaissance missions. This seems to have been in a form of diving bell, and was depicted in a 16th-century Islamic painting.

According to a report in Opusculum Taisnieri published in 1562, two Greeks submerged and surfaced in the river Tagus near the City of Toledo, several times, in the presence of The Holy Roman Emperor Charles V, without getting wet and the flame they carried in their hands remained still alight.

Before the submarines came into existence, the word was 'submersible' or 'submerged' for the objects going under water. Centuries ago, the man wished to explore under water to satisfy his curiosity; surprisingly this holds good even today; only the means and the technology have advanced, but the basics have not changed. Archimedes principle, gravity and glass are still used for the purpose. Edith Widder, a writer, biologist and a deep-sea explorer said 'Exploration is the engine that drives innovation and innovation drives economic growth; so let us all go exploring.' May be about two thousand years ago, it was the glass bell in which the man was lowered; these days one is lowered from a ship to walk on the sea bed with a heavy headgear, like a helmet, for weighing down the person; the headgear has glass in the front. Two trailing hoses are attached to the headgear, one for supplying air at a pressure higher than atmospheric pressure, for breathing and the other for exhaust. One such place is in Thailand, on the way from 'Pathaya' to 'Laan Island', where a ship is anchored and the person is lowered to walk on the sea bed which is at a depth of about 7 to 8 meters, that is the maximum pressure which normal human being can withstand. The heart patients are advised not to venture into this sport.

MODERN DAY SUBMERSIBLES

Modern day submersible is a small vehicle designed to operate underwater. The term submersible can be defined, to differentiate from other underwater vehicles known as submarines. A submarine is a fully autonomous craft, capable of renewing its own power and breathing air, whereas a submersible is usually supported by a surface vessel, platform, shore team or sometimes a larger submarine. In common usage by the general public, however, the word 'submarine' may be used to describe a craft that is by the 'technical definition' actually is a 'submersible'. There are many types of submersibles, including both crewed and unmanned, otherwise known as remotely operated vehicles or ROVs. These are also called as Deep Sea Rescue Vehicles or DSRVs. Such remotely operated vehicles are attached by a tether (a thick cable providing power and communications) to a control center on a ship. Operators on the ship see video images sent back from the robot and may control its propellers and manipulator arm. The wreck of the Titanic was explored by such a vehicle, as well as by a manned vessel. Some submersibles have been able to dive to great depths of over 10 kilometers. Single atmosphere, manned submersible, has a pressurized hull and the occupants are at the standard atmospheric pressure. This requires the hull to be capable of withstanding the high pressure from the water outside which is many times greater than the internal pressure. Submersibles have many uses worldwide, such as tourism, oceanography, underwater archaeology, ocean exploration, adventure, equipment maintenance/recovery or underwater videography.

Mass-produced underwater vehicles

Autonomous underwater vehicles are essential for tasks such as exploring the seabed in search of oil or minerals. Fraunhofer researchers have designed the first robust, lightweight and powerful vehicle intended for series production. There has never been so much human activity in the depths of the oceans. Several thousand meters below the surface, oil companies are prospecting for new deposits and deep-sea mining companies are looking for valuable mineral resources. Then there are the thousands of kilometers of pipelines and submarine cables that need regular maintenance. Not to mention the marine scientists who would like to be able to use robust devices to survey large areas of the ocean floor. All these applications mean there is a growing demand for underwater exploration vehicles. To meet this demand, researchers at the Fraunhofer Institute for Optronics, System Technologies and Image Exploitation IOSB in Ilmenau and Karlsruhe have designed a powerful autonomous underwater vehicle (AUV) capable of being manufactured in large numbers. Companies have been using AUVs for many years in deep-sea exploration missions. These untethered vehicles glide independently through the water collecting observation data, and make their own way back to the research vessel. Up to now, these have primarily been custom-built and very expensive. They have complicated structures, which makes them relatively difficult to handle by the crew on board the research vessel; for instance, accessing the batteries in order to replace them. It takes one hour to read the many terabytes of observation data out of the AUV's onboard processor. What's more, many of these vehicles are so heavy that only specially trained operators can place them in the water using the ship's winch. The IOSB's AUV overcomes all of these problems. The vehicle called DEDAVE (Deep Diving AUV for Exploration) bears a certain resemblance to the space shuttle. The research team has fitted it out with technologies not normally found in AUVs to date. To avoid the typical mess of cables, which was often a source of faults, they installed a CAN bus system like those found in every modern car. It consists of a slim cable to which all control devices and electric motors can be connected. The advantage of having so few cables and connectors is that faults are avoided. New modules, sensors or test devices can also be connected quickly and easily to the standardized CAN bus. Batteries and data storage devices are held in place by a tough but simple latch mechanism, allowing them to be removed with a minimum of effort. There is no longer any need to download data from the processor. One of the strengths of the lightweight, 3.5-meter-long underwater vehicle is that it takes up very little space. Aboard a ship, AUVs are stored in standard shipping containers, which usually offer only enough room for one vehicle. "We, on the other hand, can fit four AUVs into the same container," says the lead researcher. "The advantage of having four vehicles available is that larger than usual areas of the ocean can be surveyed in far less time." Despite their small size, the AUVs still provide plenty of additional carrying space. The payload bay measures approximately one meter in length, which is sufficient for installing several different sensors for capturing ocean floor survey data. The underwater vehicle is powered by eight batteries, each weighing 15 kilograms. A fast-release latch mechanism enables them to be removed and replaced with little effort. A fully charged battery holds enough power for up to 20 hours' travel. The software for the sophisticated battery management system was specially developed by researchers at the Fraunhofer Institute for Silicon Technology ISIT in Itzehoe. In collaboration with the GEOMAR Helmholtz Center for Ocean Research, Kiel, and a Spanish research center, DEDAVE will go through deep sea testing off the coast of Gran Canaria in the coming weeks. DEDAVE is the world's first autonomous underwater vehicle to be developed from the outset with a view to series production. It will be manufactured by a company to be specifically created for this purpose as a spin-off from the IOSB in the first half of 2016.

SUBMARINE PROPULSION—MUSCLE POWER

'Muscle Power' coupled with some mechanical advantage continued to be used, for almost three centuries, to propel the early submarines. In 1578/1580, the first published description for a submarine came, from WILLIAM BOURNE, an English innkeeper and scientific dilettante. Bourne first offered a lucid description of why a ship floats; by displacing its weight of water, and then described a mechanism by which, it is possible to make a ship or boat that may go under water, and also to come up again. The following is the drawing purported to be Bourne's scheme: leather-wrapped pads, which can be screwed in toward the centerline to create a flooded chamber, and screwed out to expel the water and seal the opening.

The first successful submarine was built in 1620 by Cornelius Jacobszoon Drebbel, a Dutchman in the service of James I; it may have been based on Bourne's design. According to accounts, some of which may have been written by people who actually saw the submarine, it was a decked-over rowboat, propelled by twelve oarsmen, which made a submerged journey down the Thames River at a depth of about fifteen feet. It was propelled by oars. The precise nature of the submarine type is a matter of some controversy; some claim that it was merely a bell towed by a boat. Two improved types were tested in the Thames between 1620 and 1624. There are no credible illustrations of Drebbel's boat, and no credible explanations of how it worked. Best guess is that the boat was designed to have almost-neutral buoyancy, floating just awash, with a downward-sloping foredeck to act as a sort of diving plane. The boat would be driven under the surface by forward momentum, just as are most modern submarines. When the rowers stopped rowing, the boat would slowly rise.

In the year 1634, French priest MARIN MERSENNE theorized that a submarine should be made of copper, cylindrical in shape to better withstand pressure and with pointed ends both for streamlining and to permit reversing course without having to turn around. For every foot of depth, water pressure increases about half a pound per square inch (PSI). Though the first submersible vehicles were tools for exploring under water, it did not take long for inventors to recognize their military potential. The strategic advantages of submarines were set out by Bishop John Wilkins of Chester in Mathematical Magick in 1648; traditional mechanical devices are discussed such as the balance, the lever, the wheel or pulley and the block and tackle, the wedge, and the screw. The possibilities are considered to improve the design of the submarine built by Drebbel.

Year 1653 saw the 72-foot-long "Rotterdam Boat," designed by a Frenchman (named DE SON) it was probably the first underwater vessel specifically built (by the Belgians) to attack an enemy (the English Navy). This almost submarine, a semi-submerged ram, was supposed to sneak up unobserved and punch a hole in an enemy ship. The designer boasted that it could cross the English Channel and back in a day, and sink a hundred ships along the way.

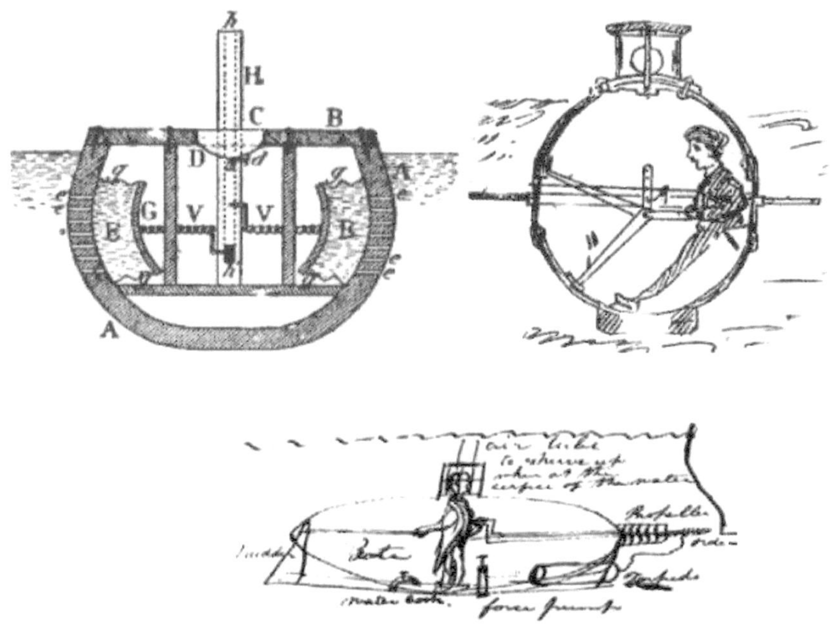

Portraits generated from the description by WILLIAM BOURNE

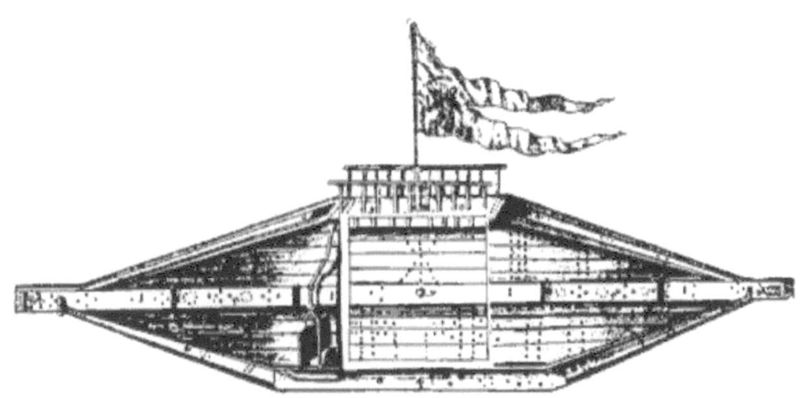

"Rotterdam Boat." Propulsion: a spring-driven clockwork device to turn a central paddle wheel. The device was so underpowered that,

when the boat was launched, it went – literally – nowhere.

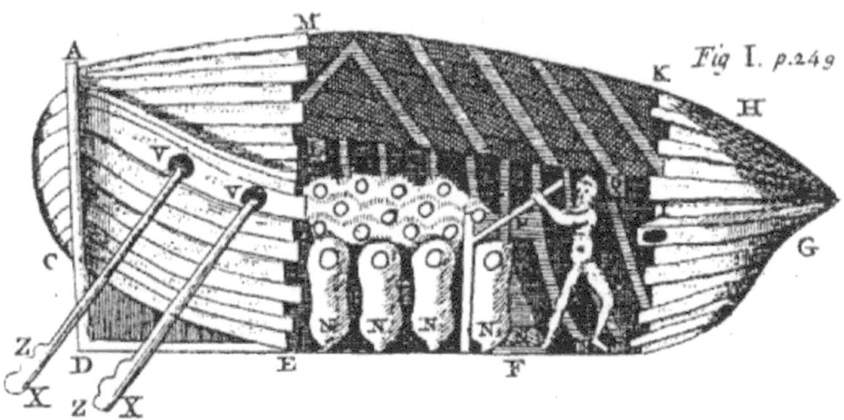

1680 -There is no evidence that Italian GIOVANNI BORELLI ever built a submarine, but this illustration continues to appear in books and magazines – in several variations – as if it was a real boat, sometimes erroneously linked with Drebbel's or Symons's efforts. Borelli did understand the basic principle of volume vs. weight (displacement), but he illustrated a totally impractical ballast system by which weight would be increased or diminished by allowing a bank of goatskin bags to fill with water, then by squeezing the water out to rise again.

In 1696, DENIS PAPIN, a professor of mathematics built two submarines. He used an air pump to balance internal pressure with external water pressure, thus controlling buoyancy, through the in-and-out flow of water into the hull. For propulsion, he used sails on the surface and oars underwater. Papin featured "certain holes" through which the operator might "touch enemy vessels and ruin them in sundry ways."

By the mid-18th century, over a dozen patents for submarines/submersible boats had been granted in England. In the year 1729, an English house-carpenter NATHANIEL SYMONS created a one-man expanding/contracting sinking boat, there was no locomotion, it was, as a sort of public entertainment. Sealed up inside, in front of a crowd of spectators, he cranked the two parts of his telescopic hull together, spent forty-five minutes underwater, then expanded the hull, rose to the surface, and passed the hat. One man gave him a coin.

In the year 1773, Wagon-maker J. DAY, another Englishman, built a small submarine with detachable ballast stones, hung around the outside, with ring bolts, which could be released from inside. This worked quite well in shallow water. Encouraged by a professional gambler, he built a bigger boat; they would take bets on how long he could remain underwater, further out in the deep-water harbor. Surrounded by ships filled with bettors, they hung some stones; the boat wallowed awash, but would not go under. They hung some more stones. The boat sank, like a rock and would have collapsed long before the ballast could be released.

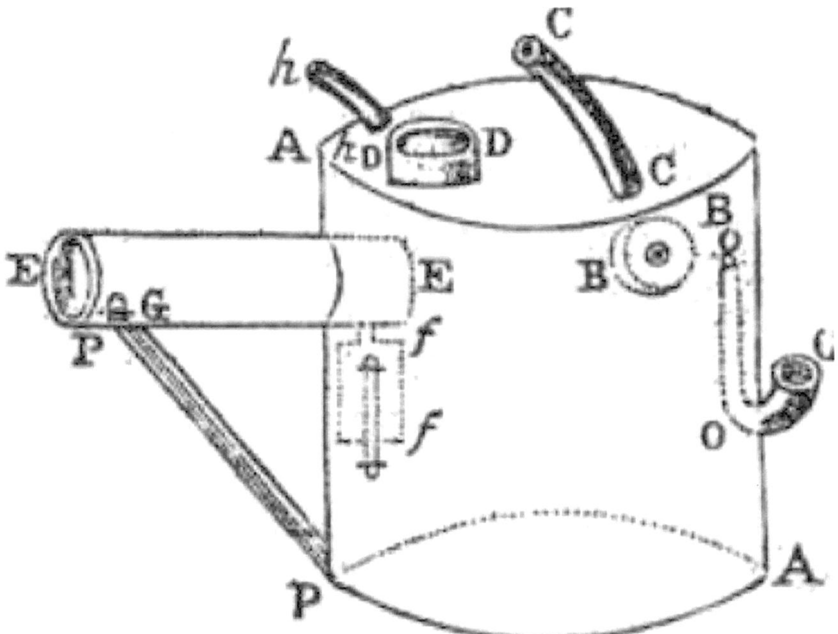

Picture from the description -- Papin tested his 1st boat, but his patron lost interest & the 2nd was never finished. Illustrations of this submarine look like a steam kettle. Papin also invented the pressure cooker.

The first military submarine was built in 1720 by carpenter Yefim Nikonov by order of Tsar Peter the Great in Russia. Nikonov armed his submarine with “fire tubes”, weapons akin to flame-throwers. The submarine was designed to approach an enemy vessel, put the ends of the “tubes” out of the water, and blow up the ship with a combustible mixture. In addition, he designed an airlock for aquanauts to come out of the submarine and to destroy the bilge of the ship. With the death of Peter I in January 1725, Nikonov lost his principal patron and the Admiralty withdrew support for the project.

The first American military submarine was Turtle in 1776, a hand-powered egg-shaped (or acorn-shaped) device designed by the American David Bushnell, to accommodate a single man.

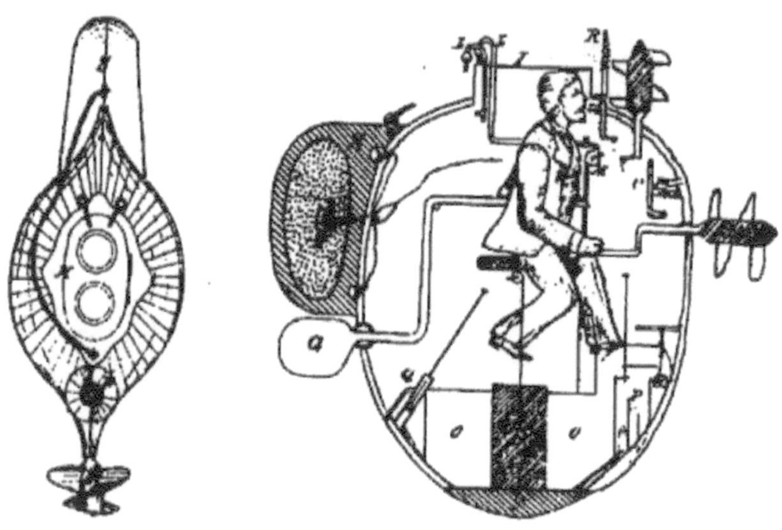

"Turtle," as drawn in 1875 from the best information the artist could gather. There are several important errors. It shows ballast tanks when there were none; it shows an Archimedes screw (helical) for locomotion instead of the propeller like the "arms of a wind mill" or a "pair of oars" described by Bushnell and others.

It was the first submarine capable of independent underwater operation and movement, and the first to use screws for propulsion., Yale graduate DAVID BUSHNELL built the first submarine to actually make an attack on an enemy warship. Dubbed the "Turtle" because it resembled a sea turtle floating vertically in the water, it was operated by Sergeant Ezra Lee. The scheme was that, it would be towed into the vicinity of the target; a foot operated valve would be opened to let in enough water to sink, the valve would be closed and it would be moved under the enemy by cranking the two propellers, one for forward and one for vertical movement. A drill would be turned by foot treadle like a spinning wheel" to drill into the hull, to attach a 150-pound keg of gunpowder with a clockwork detonator; the cranking would be done to get away. A foot-pump would be operated to get the water out of the hull and thus re-surface. In early-morning darkness on September 7, 1776, "Turtle" made an attack on a British ship in New York harbor, probably HMS Eagle. The drill may have hit an iron strap; it would not penetrate the hull. Contrary to most reports, the HMS Eagle of 1776 had not been fitted with a copper-sheathed bottom, so the drill did not penetrate. Lee became disoriented, soon bobbed to the surface and was spotted by a lookout. He managed to get away.

In the year 1797, ROBERT FULTON, a marginal American artist but increasingly successful inventor living in Paris, offered to build a submarine to be used against France's British enemy. Finally, in the year 1800, after protracted delays and several changes in government, Fulton was encouraged enough to build the submarine he called "Nautilus." It also had a sail for use on the surface and so was the first known use of dual propulsion on a submarine. It proved capable of using mines to destroy two warships during demonstrations. He made a number of successful dives, to depths of 25 feet and for times as long as six hours with ventilation provided by a tube to the surface. "Nautilus" was essentially an elongated "Turtle" with a larger propeller and mast and sail for use on the surface. In trials, "Nautilus" achieved a maximum sustained underwater speed of four knots. Fulton, given the rank of rear admiral, made several attempts to attack English ships, which saw him coming and moved out of the way. Relationships with the French government deteriorated; a new Minister of Marine is reported to have said, "Go, sir. Your invention is fine for the Algerians or corsairs, but be advised that France has not yet abandoned the Ocean." Fulton broke up "Nautilus" and sold the metal for scrap. He proposed an improved version, but, there are reports to the contrary, that it was never built. The name "Nautilus" was immortalized by Jules Verne in his 1870 novel, "20,000 Leagues under the Sea" and was given to several U. S. Navy boats – including the world's first nuclear-powered submarine, in the 1954 USS Nautilus.

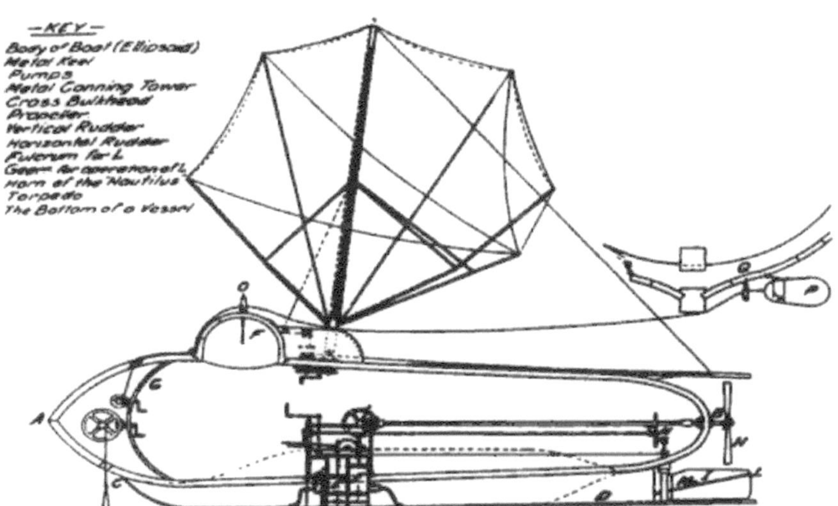

This most commonly reproduced "Nautilus" was drawn two years before the submarine was built; Fulton added a deck and made a number of un-documented changes in the finished product.

Illustrations, which show "Nautilus" with the hull-form and sail rig of a surface sailboat represent the never-built "improved" version.

Year 1812-1815-operator with one hand on a tiller, the other on a crank to turn the propeller and drill bit. (From the note books, a design attributed to SILAS CLOWDEN HALSEY). A "water cock" and a "force pump" at the bottom of the boat and an "air tube to shove up when at the surface of the water." A "torpedo" is attached by a line to the drill.

During the War of 1812, in 1814 Silas Halsey lost his life while using a submarine in an unsuccessful attack on a British warship stationed in New London harbor.

In 1834 a Russian naval designer Karl Shilder built and tested an all-metal submarine in Saint Petersburg. His submarine was equipped by 6 Congreve rockets

The Submarino Hipopótamo was the first submarine in South America built and tested in Ecuador on September 18, 1837. It was designed by Jose Rodriguez Lavandera, who successfully crossed the Guayas River in Guayaquil accompanied by Jose Quevedo. Rodriguez Lavandera had enrolled in the Ecuadorian Navy in 1823, becoming a Lieutenant by 1830. The Hipopotamo crossed the Guayas on two more occasions, but it was abandoned, because of lack of funding and interest from the government.

1850-The German port of Kiel was under blockade by the Danish Navy, and Prussian army corporal WILHELM BAUER persuaded a shipbuilder to construct his design for a blockade-breaking submarine, which he called "Brandtaucher," (Incendiary Diver). The boat was made of riveted sheet iron, about the size and shape of a small sperm whale; propulsion was by a two-man-power treadmill which drove a propeller and a third crewmember steered. Buoyancy was controlled by ballast tanks, and trim was adjusted by moving a sliding weight along an iron rod.

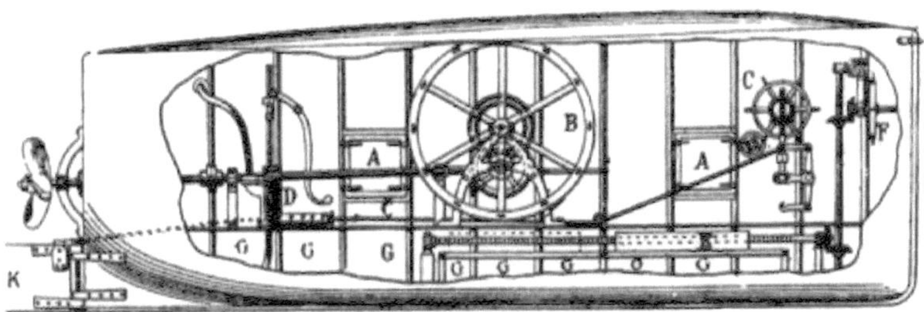

"Brandtaucher" was recovered in 1887 and is now on display in Dresden.

Year 1863- The propulsion remained essentially as 'Muscle Power' Hunley's New Orleans consortium built a second, slightly improved submarine, which may have been called "American Diver." McClintock spent some time and money trying to replace hand cranking with some sort of electrical motor, but without success.

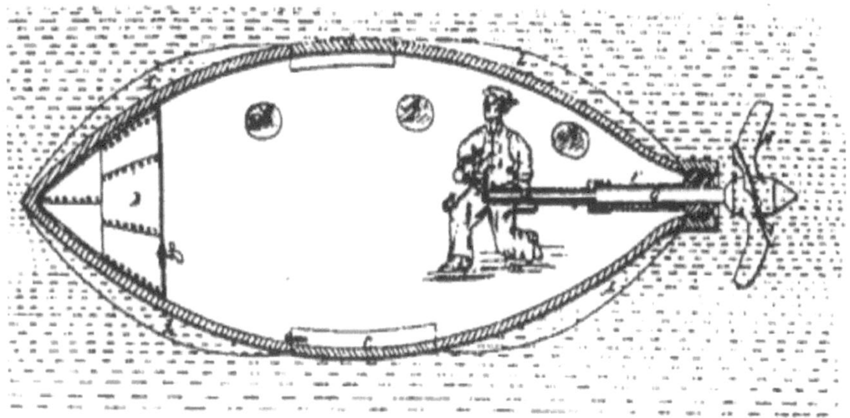

Phillips was granted a 1852 patent for a "Steering Submarine Propeller." The innovation: steering (as well as up-and-down movement) was controlled by a hand-cranked propeller on a swivel joint.

The "Flach" was commissioned in 1865 by the Chilean government during the war between Chile and Peru against Spain (1864–1866). It was built by the German engineer Karl Flach. The submarine sank during tests in Valparaiso bay, on May 3, 1866, with the entire eleven-man crew. During the War of the Pacific in 1879, the Peruvian government commissioned and built a submarine, the Toro Submarino. It never saw military action and was scuttled after Peru's defeat to prevent its capture by the enemy.

Photo # NH 58769 Cutaway drawings of the Confederate submarine H.L. Hunley

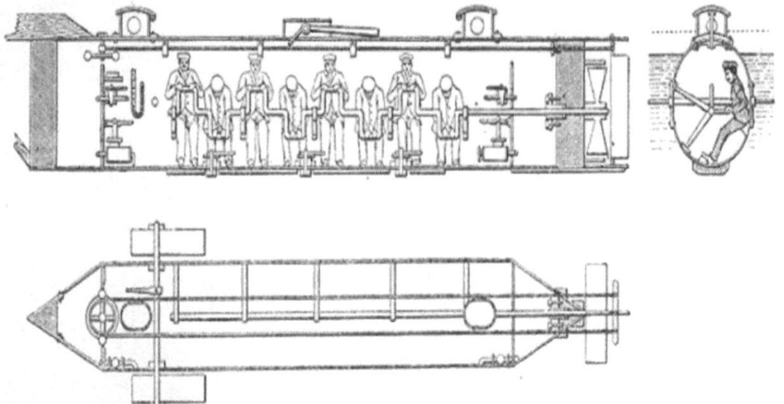

Fig. 175 à 177. — Le *David* de Hunley reconstitué d'après les dessins de M. William-A. Alexander (1863).

The propulsion still remained 'Muscle Power' the cross-section) clearly shows the tight working space inside.

The latter half of 19th century was the era of Hunley and warfare. The Hunley's first crew was lost in August, 1863, the second in October, 1863 and the third crew was lost in February, 1864, after attacking the USS Housatonic. The Hunley was the first true submarine to successfully attack and sink a ship in combat.

Following the Hunley's, submarines, the submarines were not used in warfare until 1914. The submarines created after the Hunleys had many of the same features as the Hunleys.

MECHANICAL POWER REPLACES MUSCLE POWER

The first attempt to propel without muscle power was in 1863. A French team of CHARLES BURN and SIMON BOURGEOIS launched "Le Plongeur" (The Diver). It was 140 feet long, 20 feet wide, displacing 400 tons. The engines were run by 180 PSI compressed air stored in tanks throughout the boat. Method of operation was to fill ballast tanks just enough to achieve neutral buoyancy, and then make adjustments with cylinders that could be run in and out of the hull to vary the volume (Bourne's concept). The boat was too unstable; the movement of a crew member could send her into radical gyrations.

In 1864, WILHELM BAUER proposed that submarines were to be powered by a visionary, but not yet practical; his proposal was to use an internal combustion engine. Overall, he was to spend twenty-five years developing, or at least proposing submarines on behalf of six nations – Germany, Austria, England, the United States, Russia, and France.

After a) McClintock spent some time and money trying to replace hand cranking with some sort of electrical motor in 1863, b) the French team of CHARLES BURN and SIMON BOURGEOIS tried Engines run by 180 PSI compressed air stored in tanks and c) WILHELM BAUER proposed that submarines be powered by Internal combustion engine, the Irish inventor John Philip Holland built a model submarine in 1876 and a full scale one in 1878. Holland found sponsorship with the Fenians, a group of Irish revolutionaries, and built a small prototype submarine, "Holland No. 1" to test out his theories, including the use of a Gasoline engine; though a number of theories were unsuccessful. In 1896 he designed his Holland Type VI submarine, which for the first time, made use of internal combustion engine power on the surface and electric battery power for submerged operations. Launched on 17 May 1897 at Lewis Nixon's Crescent Shipyard in Elizabeth, New Jersey, the Holland VI was, eventually purchased by the United States Navy on 11 April 1900, becoming the United States Navy's first commissioned submarine and renamed USS Holland. A prototype version of the A-class submarine (Fulton) was developed soon after, at Crescent under the supervision of naval architect Arthur Leopold Busch for the newly reorganized Electric Boat Company. The Fulton was never commissioned by the United States Navy and was sold to the Imperial Russian Navy in 1905. Many countries became interested in Holland's (weapons) product and purchased "the rights" to build them during this time period. The Holland Torpedo Boat Company/Electric Boat Company became General Dynamics. This "Cold War" progeny became, arguably the builder of the world's most technologically advanced submarines to this day.

The construction of the A-class boats soon followed the prototype (Fulton). The submarines were built at two different shipyards on both coasts of the United States; Union Iron Works/Mare Island Naval Shipyard and Crescent Shipyard. In 1902, Holland received U.S. Patent 708,553 for his relentless pursuit to perfect the modern submarine craft. Some of his vessels were purchased by the United States and other "technologically advanced" nations such as the United Kingdom, the Imperial Russian Navy, the Royal Netherlands Navy and the Imperial Japanese Navy. Mr. Holland was no longer in control of his company at this point, as others were formally engaged in transactions with many other foreign nations around the world at this time. The Type VII design was also adopted by the Royal Navy as well (with Holland's input, as the Holland class submarine, including Britain's Holland #1).

Holland design included many of the features of modern submarines. This vessel was displacing 63 tons, had an overall length of 53 feet (16.2m) and a submerged speed of 5 knots (2.6 m/s). Its range was 1500 nautical miles (2,778km). There was an aft propeller with control vanes for steering. Its general proportions were not that different from the latter, more sophisticated Albacore. Its profile was streamlined, a small rounded nose increasing to the full cross-section followed by a tapering tail. Its length to breadth ratio of 5.05 was not far from the optimum. Because of its need of oxygen for combustion, the petrol engine could not be used when the submarine was submerged. However, when the craft was on the surface and the hatch open, the engine could operate a generator to recharge batteries and it could also then propel the ship. Then it suffered from water entering through the open hatch due to the low freeboard. This prompted the addition of a larger raised conning tower in later designs. A single propeller was arranged at the tail with control vanes, also a rudder and elevators. The trial was successful enough to encourage building a larger, more warlike boat.

In the year 1879, Anglican Reverend GEORGE W. GARRETT tested the Steam-powered "Resurgam:" Steam from a boiler was used for surface operations and steam stored in pressurized tanks for submerged operations. The boat NORDENFELDT passed initial trials, but sank while undertow (rediscovered in 1996). Out of funds but not undeterred, Garrett took his ideas to a wealthy Swedish arms manufacturer.

In the year1885, French designer CLAUDE GOUBET built a Battery-operated submarine; It was too awkward and unstable to be successful. He followed up in 1889 with "Goubet II", also small, electric, but not effective.

In the year 1885, the American JOSIAH H.L.TUCK demonstrated "Peacemaker", it was powered by A chemical (fireless) boiler; 1500 pounds of caustic soda provided five hours endurance. Tuck's inventing days ended when relatives, noting that he had squandered most of a significant fortune, had him committed to an asylum for the insane.

Photograph of Holland, 1900

In the Year 1888, GUSTAVE ZEDE built "Gymnote" for the French Navy A Battery-powered boat but limited by the lack of any method for recharging the batteries while at sea.

With a new Administration in office, the U. S. Congress appropriated $200,000 for an "experimental submarine" and the Navy announced a new competition. There were three bidders: HOLLAND, SIMON LAKE and GEORGE C. BAKER,. Holland and Lake submitted proposals. The politically well- connected Baker actually had a submarine, which he was demonstrating on Lake Michigan. A novel feature was a clutch between the steam engine and an electric motor, allowed the motor to function as a dynamo, to recharge the batteries for submerged running.

When Holland's design once again won the competition, Baker complained to his friends in Washington. The whole business seems to have been put on "hold. The scheme that Simon Lake submitted included a set of wheels by which the boat could run along the bottom. This boat used a gasoline engine for both surface and submerged running (drawing air from the surface through breathing tubes.) He tested this theory in 1894 with small wooden "test vehicle" dubbed "Argonaut Jr." and financed by relatives.

Finally, in 1895 John P. Holland Torpedo Boat Company was awarded the $200,000 to build a 85-foot, 15-knot, steam-powered submarine to be called "Plunger." Holland was not fully satisfied and pleased, because he didn't like the imposition of a steam engine, as well some changes the Navy insisted upon. Congress was thrilled, and immediately authorized two more submarines of the Plunger type at $175,000 each. "Plunger," launched in 1897, failed before ever leaving the dock. The temperature in the fire room reached 137 degrees at only 2/3 rated output. As one of Holland's employees was later to testify, "They forced us to put steam in the "Plunger" against Mr. Holland's advice. Even before "Plunger" had failed, Holland began construction of a new, smaller (54 feet), slower (7 knots), gasoline-powered boat, "Holland VI."

In September, 1898, SIMON LAKE'S 36-foot "Argonaut I" made an open-ocean passage from Norfolk, VA, to Sandy Hook, NJ, prompting Jules Verne to send Lake a cable: "The conspicuous success of submarine navigation in the United States will push on under-water navigation all over the world . . ." and "The next war may be largely a contest between submarine boats." Simon Lake independently arrived at the solution, combining the petrol engine and the battery. His vessel, Protector, was launched in 1902. His greatest contribution was to overcome the problem of sight because submarines when submerged were blind. With the aid of a professor of optics, from John Hopkins, they constructed the forerunner of the periscope. Lake's submarines were sold to Russia and Austria. Later they were bought by the US Navy, as were those of Holland.

In the same year, MAXIME LAUBEUF'S "Narval," 188-feet, 136-tons, began life with steam power but was switched to a Diesel engine.

It was 19th century and the era of Holland's. The turn of the century, marked a pivotal time in the development of submarines, with a number of important technologies making their debut and the widespread adoption and fielding of submarines by a number of nations. Diesel Electric propulsion would become the dominant power system and instruments such as the periscope would become standardized design. Large numbers of experiments were done by countries on effective tactics and weapons for submarines; all of which would culminate in making a large impact on 20th century warfare. By October, 1900, the British had five Holland submarines on order, but not until the senior naval leadership had wrestled with a moral dilemma. They, like many others through the years, believed that covert warfare was, basically, illegal. Gentlemen fought each other face to face, wearing easily recognized uniforms. The Navy agreed to proceed with caution, primarily to "test the value of the submarine as a weapon in the hands of our enemies." However, Rear Admiral A. K. Wilson assured himself of a certain immortality by declaring that the submarine was "underhand, unfair, and damned Un-English." The government, he wrote, should "treat all submarines as pirates in wartime and hang all crews."

In the year 1904, John P. Holland was squeezed out of management and was increasingly ignored; he resigned from Electric Boat and formed John P. Holland's Submarine Boat Company. He sold plans for two larger, improved submarines, to be built in Japan, under the supervision of a Holland associate; one achieved a remarkable underwater speed of 16 knots, about twice that of the five earlier models of Holland submarines in Japan. Mr. Holland solicited business from around the world, but quickly discovered that all of his patents were controlled by Electric Boat, a fact of which the company made certain that all potential customers were aware. He tried to interest the U. S. Navy in a new, fast hull design; he tested the same in an experimental tank at the Washington navy Yard. It promised submerged speeds as high as 22 knots. The Navy offered the opinion that it would be too hazardous for submarines to go faster than 6 knots underwater. Electric Boat sued John P. Holland for breach of contract, for unethical conduct, and even for using the name "Holland." The suits were eventually dismissed by the courts. Nevertheless, Mr. Holland's business never recovered.

ELECTRIC POWER

A reliable means of propulsion for the submerged vessel was only made possible in the 1880s with the advent of the necessary electric battery technology. The first electrically powered boats were built by Stefan Drzewiecki in Russia, James Franklin Waddington and the team of James Ash and Andrew Campbell in England, Dupuy de Lôme and Gustave Zédé in France and Isaac Peral in Spain.

In 1884, Polish-Russian naval engineer Stefan Drzewiecki converted 2 mechanical submarines, installed on each one 1 hp engine with the new source of energy - batteries. On tests submarine went under the water against the flow of the Neva River at a rate of 4 knots. It was the first submarine in the world with electric propulsion. Ash and Campbell constructed their craft, the Nautilus, in 1886. It was 60 feet (18 m) long with a 9.7 kW (13 hp) engine powered by 52 batteries. It was an advanced design for the time, but became stuck in the mud during trials and was discontinued. Waddington's Porpoise vessel showed more promise. Waddington had formerly worked in the shipyard in which Garrett had been active. Waddignton's vessel was similar in size to the Resurgam and its propulsion system used 45 accumulator cells with a capacity of 660 ampere hours each. These were coupled in series to a motor driving a propeller at about 750 rpm, giving the ship a sustained speed of 13 km/h (8 mph) for at least 8 hours. The boat was armed with two externally mounted torpedoes as well as a mine torpedo that could be detonated electronically. Although the boat performed well at trials, Waddington was unable to attract further contracts and went bankrupt.

In France, early electric boats Goubet I and Goubet II were built by the civil engineer, Claude Goubet. These boats were also unsuccessful, but they inspired the renowned naval architect Dupuy de Lôme to begin work on his submarine – an advanced electric-powered submarine almost 20 meters long. He didn't live to see his design constructed, but the craft was completed by Gustave Zédé in 1888 and named the Gymnote. It was one of the first truly successful electrically powered submarines, and was equipped with an early periscope and an electric gyrocompass for navigation. It completed over 2,000 successful dives using a 204-cell battery. Although the Gymnote was scrapped for its limited range, its side hydroplanes became the standard for future submarine designs.

The Peral Submarine, constructed by Isaac Peral, was launched by the Spanish Navy in the same year, 1888. It had three Schwarzkopf torpedoes 14 in (360 mm) and one torpedo tube in bow, new air systems, hull shape, propeller, and cruciform external controls anticipating much later designs. Peral was an all-electrical powered submarine. After two years of trials the project was scrapped by naval officialdom that cited, among other reasons, concerns over the range permitted by its batteries.

Many more designs were built at this time by various inventors, but submarines were not put into service by navies until the turn of the 20th century.

In 1906, U-1, the first German "U-Boat" (a short form for unterzeeboot), was launched. This modified "Karp" was 139 feet long, displaced 239 tons. It had a surface speed of 11 knots, a submerged speed of 9 knots, and a range of two thousand miles. In the year 1902, Krupp, in Germany, worked on a larger, improved design – the "Karp" class – powered by gasoline engine on the surface, with an onboard battery recharging system. Russia ordered three submarines. The German Navy ordered one of them, but asked for a Kerosene rather than gasoline engine.

Year 1911, Thanks in large part to the efforts of a 26-year old Navy lieutenant, Chester Nimitz, who by this time had commanded three U. S. submarines, the obnoxious and dangerous gasoline engine were replaced by diesels, beginning with Nimitz' fourth submarine command, "Skipjack." Petrol fumes present within the hull had all the chances of a catastrophic explosion. The problem was solved by the invention of the compression ignition engine by a German engineer, Rudolf Diesel; this had no electric spark and was running on cheaper, much less volatile fuel oil. The diesel engine, as it was called, was more efficient and more economical, giving greater range. The fumes were less toxic and less volatile.

Henceforward it became the accepted main source of power until the advent of nuclear power, although the development of other AIP (Air Independent Propulsion) Systems have continued till date; there was some slowdown during the development of 'Nuclear Propulsion' which was termed as the best AIP.

Year 1912, Germany began to get serious about submarines with the "30s" series – U-31 to U-41. These diesel-powered boats displaced 685 tons, carried six torpedoes and one 88mm deck gun had a surface speed of 16.4 knots, submerged 9.7 knots – and a maximum range of 7,800 miles at 8 knots.

SUBMARINE WARFARE DURING FIRST AND SECOND WORLD WARS

Shortly before the outbreak of World War I, submarines were employed by the Italian Regia Marina during the Italo-Turkish War without seeing any naval action, and by the Greek Navy during the Balkan Wars, where notably the French-built Delfin became the first such vessel to launch a torpedo against an enemy ship (albeit unsuccessfully).

The first time military submarines had significant impact on a war was in World War I. Forces such as the U-boats of Germany operated against Allied commerce (Handelskrieg). The submarine's ability to function as a practical war machine relied on new tactics, their numbers, and submarine technologies such as combination diesel/electric power system that had been developed in the preceding years. More like submersible ships than the submarines of today, submarines operated primarily on the surface using standard engines, submerging occasionally to attack under battery power. They were roughly triangular in cross-section, with a distinct keel, to control rolling while surfaced, and a distinct bow.

At the start of the war, Germany had 48 submarines in service or under construction, with 29 operational. These included vessels of the diesel-engine U-19 class with the range (5,000 miles) and speed (eight knots) to operate effectively around the entire British coast. Initially, Germany followed the international "Prize Rules", which required a ship's crew to be allowed to leave before sinking their ship. The U-boats saw action in the First Battle of the Atlantic.

France had 62 submarines at the beginning of the war, in 14 different classes. They operated mainly in the Mediterranean, and in the course of the war, 12 were lost. The Russians started the war with 58 submarines in service or under construction. The main class was the "Bars" with 24 boats. Twenty-four submarines were lost during the war.

The British had 77 operational submarines at the beginning of the war, with 15 under construction. The main type was the "E class", but several experimental designs were built, including the "K class", which had a reputation for bad luck, and the "M class", which had a large deck-mounted gun. The "R class" was the first boat designed to attack other submarines. British submarines operated in the Baltic, North Sea and Atlantic, as well as in the Mediterranean and Black Sea. Over 50 were lost from various causes during the war.

In August 1914, a flotilla of ten U-boats sailed from their base in Heligoland to attack Royal Navy capital ships of the British Grand Fleet, and so reduce the Grand Fleet's numerical superiority over the German High Seas Fleet. Depending more on luck than strategy, the first sortie was not a success. Only one attack was carried out, when U-15 fired a torpedo (which missed) at HMS Monarch, while two of the ten U-boats were lost. The SM U-9 had better luck. On 22 September 1914 while patrolling the Broad Fourteens, a region of the southern North Sea, U-9 found three obsolescent British Cressy-class armored cruisers (HMS Aboukir, Hogue, and Cressy), which were assigned to prevent German surface vessels from entering the eastern end of the English Channel. The U-9 fired all six of its torpedoes, reloading while submerged, and sank the three cruisers in less than an hour.

After the British ordered transport ships to act as auxiliary cruisers, the German navy adopted unrestricted submarine warfare; generally giving no warning of an attack. During the war, 360 submarines were built, but 178 were lost. The rest were surrendered at the end of the war. A German U-boat sunk RMS Lusitania and is often cited among the reasons for the entry of the United States into the war. World War I accelerated progress in the design of submarines. The German navy developed its long, heavy, long ranging, diesel powered Unterseeboote or U-boats, as they were known through two world wars. They were essentially surface running ships with fine bows and usually mounting a gun, a large bridge fin and superstructure, under slung twin propellers, many excrescences and innumerable flooding and venting holes for rapid surfacing and submerging. No attention was paid to underwater performance. The diesel engine and the electric battery remained the main power source for submarines. Most of the advances were made in armament, speed, periscopes, snorkel, radar, sonar, hull construction, hydrophones etc.

The art of submarine warfare was barely a dozen years old. No nation had developed any method for detecting submarines, or attacking them if found. Thus began development of the depth charge, which claimed its first victim in March 1916. However, overall, these depth charges were not very effective unless exploding quite close to the U-boat. The main benefit was psychological. It was Germany's use of the U-boat in World War I that demonstrated the vital role the submarine would play in the next global conflict. As the motor had driven the horse, from the road, so had the submarine driven the battleship from the sea! The development of submarine-locating devices began early in the war with hydrophones (a directional microphone in the water) to listen for the sounds of propellers, it took too long to be of much use in this war. This is an echo-ranging system, (the British dubbed it ASDIC – which apparently stands for nothing in particular – but now known universally as SONAR, which stands for "Sound Navigation and Ranging.") which sent out an audible "ping" and measuring the echo return, an operator can determine the range and bearing of a submarine.

Year 1932, U. S. Navy opened a competition for the development of a lightweight diesel engine, more suitable to submarines than any currently in production. While the number of engines (which might be purchased for submarines) was too small to justify the investment, there was a large commercial market waiting in the wings like the railroad. The Japanese were more serious about submarine aircraft carriers than any other navy; they built their first, the 2,243 tones, 320-foot I-5, in 1932. It was equipped with one floatplane. In the next 12 years, they built 28 more, in ever-increasing sizes. In 1939,the British developed an on-board escape system, Submarine Rescue Chamber, whereby sailors waiting their turn to go out through a pressure-modulated airlock (and chest-deep in water) would be able to breathe through individual oxygen masks, permanently stored in the fore and aft torpedo rooms.

In the same year, Dr. Ross Gunn of the U. S. Naval Research Laboratory suggested that "fission chambers" using an isotope of uranium, U-235, could be used to power submarines. In a "Saturday Evening Post" article, a year later, a science writer noted that one pound of U-235 has the equivalent energy of 5 million pounds of coal: "A five pound lump of only 10 to 50 percent purity would be sufficient to drive ocean liners and submarines back and forth across the seven seas without refueling for months."

Germany entered World War II with a more refined version of submersible, still not a genuine submarine, but with its characteristic emphasis on surface performance.

After the attack on Pearl Harbor, many of the U.S. Navy's front-line Pacific Fleet surface ships were destroyed or severely damaged. The submarines survived the attack and carried the war to the enemy. Lacking support vessels, the submarines were asked to independently hunt and destroy Japanese ships and submarines. They did so very effectively.

During World War II, the submarine force was the most effective anti-ship and anti-submarine weapon in the entire American arsenal. Submarines, though only about 2 percent of the U.S. Navy destroyed over 30 percent of the Japanese Navy, including 8 aircraft carriers, 1 battleship and 11 cruisers. U.S. submarines also destroyed over 60 percent of the Japanese merchant fleet, crippling Japan's ability to supply its military forces and industrial war effort. Allied submarines in the Pacific War destroyed more Japanese shipping than all other weapons combined. This feat was considerably aided by the Imperial Japanese Navy's failure to provide adequate escort forces for the nation's merchant fleet.

Whereas Japanese submarine torpedoes of the war are considered the finest, those of U.S. Navy were considered the worst. For example, the U.S. Mark 14 torpedo typically ran ten feet too deep. The faulty depth control mechanism of the Mark 14 was corrected in August 1942, but field trials for the exploders were not ordered until mid-1943, when tests in Hawaii and

Australia confirmed the flaws. In addition, the Mark 14 sometimes suffered circular runs, which sank at least one U.S. submarine. Fully operational Mark 14 torpedoes were not put into service until September 1943. The Mark 15 torpedo used by U.S. surface combatants had the same problem and was not fixed until late 1943. One attempt to correct the problems resulted in a wake less, electric torpedo being placed in submarine service. Given the prevalence of circular runs, there were probably other losses among boats which simply disappeared.

During World War II, 314 submarines served in the United States Navy, of which nearly 260 were deployed to the Pacific. On December 7, 1941, 111 boats were in commission and 203 submarines were commissioned during the war. During the war, 52 US submarines were lost to all causes, with 48 directly due to hostilities. 3,505 sailors were lost, the highest percentage killed in action, of any US service arm in World War II. U.S. submarines sank 1,560 enemy vessels, a total tonnage of 5.3 million tons, including 8 aircraft carriers, a battleship, three heavy cruisers, and over 200 other warships, and damaged several other ships including the battleships. In addition, the Japanese merchant marine lost 16,200 sailors and 53,400 wounded, of some 122,000 at the start of the war, due to submarines.

Japan had the most varied fleet of submarines of World War II, including manned torpedoes, midget submarines, medium-range submarines, purpose-built supply submarines (many for use by the Army), long-range fleet submarines (many of which carried an aircraft), submarines with the highest submerged speeds of the conflict and submarines that could carry multiple aircraft (WWII's largest submarine, the Sentoku I-400). These submarines were also equipped with the most advanced torpedo of the conflict, the oxygen-propelled Type 95 (what U.S. historian Samuel E. Morison postwar called "Long Lance").

Overall, despite their technical prowess, Japanese submarines, having been incorporated into the Imperial Navy's war plan, in contrast to Germany's war plan, they were relatively unsuccessful. Japanese submarines were primarily used in the offensive roles against warships, which were fast, maneuverable and well-defended compared to merchant ships. In 1942, Japanese submarines sank two fleet aircraft carriers, one cruiser, and several destroyers and other warships, and damaged many others, including two battleships. They were not able to sustain these results afterward, as Allied fleets were reinforced and became better organized. By the end of the war, submarines were instead often used to transport supplies to island garrisons. During the war, Japan managed to sink about 1 million tons of merchant shipping (184 ships), compared to 1.5 million tons for Great Britain (493 ships), 4.65 million tons for the U.S. (1,079 ships) and 14.3 million tons for Germany (2,840 ships).

Early models were not very maneuverable under water, could not dive very deep, and lacked radar. Later in the war, units that were fitted with radar were in some instances sunk due to the ability of U.S. radar sets to detect their emissions. After the war, several of Japan's most original submarines were sent to Hawaii for inspection in "Operation Road's End" before being scuttled by the U.S. Navy in 1946, when the Soviets demanded access to the submarines as well.

Italy had 116 submarines in service at the start of the war, with 24 different classes. These operated mainly in the Mediterranean theatre. Some were sent to a base at Bordeaux in Occupied France. A flotilla of several submarines also operated out of the Eritrean colonial port of Massawa.

Italian designs proved to be unsuitable for use in the Atlantic Ocean. Italian midget submarines were used in attacks against British shipping near the port of Gibraltar.

Although Germany was banned from having submarines in the Treaty of Versailles, construction started in secret during the 1930s. When this became known, the Anglo-German Naval Agreement of 1936 allowed Germany to achieve parity in submarines with Britain.

Germany started the war with only 65 submarines, with 21 at sea when war broke out. Germany soon built the largest submarine fleet during World War II. Due to the Treaty of Versailles limiting the surface navy, the rebuilding of the German surface forces had only begun in earnest a year before the outbreak of World War II. Having no hope of defeating the vastly superior Royal Navy decisively in a surface battle, the German High Command planned on fighting a campaign of "Guerre de course" (Merchant warfare), and immediately stopped all construction on capital surface ships, save the nearly completed Bismarck-class battleships and two cruisers, and switched the resources to submarines, which could be built more quickly. Though it took most of 1940 to expand production facilities and to start mass production, more than a thousand submarines were built by the end of the war.

Germany used submarines to devastating effect in World War II during the Battle of the Atlantic, attempting but ultimately failing to cut off Britain's supply routes by sinking more ships than Britain could replace. The supply lines were vital to Britain for food and industry, as well as armaments from Canada and the United States. Although the U-boats had been updated in the intervening years, the major innovation was improved communications, encrypted using the famous Enigma cipher machine. After putting to sea, the U-boats operated mostly on their own trying to find convoys in areas assigned to them by the High Command. If a convoy was found, the submarine did not attack immediately, but shadowed the convoy and radioed to the German Command to allow other submarines in the area to find the convoy. The submarines were then grouped into a larger striking force and attacked the convoy simultaneously, preferably at night while surfaced.

During the first few years of World War II, the U-boat force, scored unprecedented success with these tactics but were too few to have any decisive success. By the spring of 1943, German U-boat construction was at full capacity, but this was more than nullified by increased numbers of convoy escorts and aircraft, as well as technical advances like radar and sonar. High Frequency Direction Finding (HF/DF, known as Huff-Duff) allowed the Allies to route convoys around wolf packs when they detected radio transmissions from trailing boats. The results were devastating: from March to July of that year, over 130 U-boats were lost, 41 in May alone. Concurrent Allied losses dropped dramatically, from 750,000 tons in March to 188,000 in July. Although the Second Battle of the Atlantic would continue to the last day of the war, the U-boat arm was unable to stem the tide of personnel and supplies, paving the way for D-Day. Winston Churchill wrote the U-boat "peril" was the only thing to ever give him cause to doubt eventual Allied victory.

In the year 1943, hoping to hide existing U-boats from the increasingly devastating air patrols, Germany perfected an idea that had been kicking around for a long time: use of a breathing tube to allow running on diesel power just below the surface, thus also keeping the batteries fully charged. They dubbed it the "snorkel." It was not a perfect solution; the tube could break if the boat was going too fast; the ball-float valve at the top would close if a wave passed over, thus shifting engine suction to the interior of the boat and occasionally popping a few eardrums. The snorkel also left a visible wake, and also returned a pretty good radar blip, but it helped. On Type VII U-boats, the snorkel folded forward and was stored in a recess on the port side of the hull, while on the IX Types the recess was on the starboard side. The XXI and XXIII types both had telescopic masts that rose vertically through the conning tower close to the periscope.

Snorkels created several problems for their users. A U-boat with a snorkel raised was limited to six knots to avoid breaking the tube, and its sound-detection gear was useless with the diesel engine running. Most snorkels were equipped with automatic valves to prevent seawater from being sucked into the diesels, but when these valves slammed shut, the engines would draw air from the boat itself before shutting down, causing a partial vacuum which was extremely painful to the ears of the crew and sometimes even ruptured eardrums. This problem still exists in later model diesel submarines, but is mitigated by high-vacuum cut-off sensors that shut down the engines when the vacuum in the ship reaches a pre-set point.

While the snorkel renders a submarine far less detectable, it is not perfect. In clear weather, diesel exhaust can be seen on the surface to a distance of about three miles, while 'periscope feather' (the wave created by the snorkel or periscope moving through the water), is visible from far off in calm sea conditions. Modern radar is also capable of detecting a snorkel in calm sea conditions. Early Royal Navy ship's radar set Model 271 was shown in tests conducted in 1940 to detect the periscope tip of a submerged submarine at a distance of half a mile.

With the increasing sophistication of Allied detection and subsequent losses, German designers, began to fully realise the potential for a truly submerged boat. The Type XXI "Elektroboot" was designed to favor submerged performance, both for combat effectiveness and survival. It was the first true submersible. In response to the mounting U-boat losses in 1943, a submarine research centre was created by the Germans at Blankenburg in the Harz Mountains, with the aim of producing an operational high-speed U-boat capable of prolonged submergence. The Type 21 that resulted, represented a compromise since it was aimed at AIP using hydrogen peroxide and the Walter propulsion system, which had not then advanced beyond the experimental stage. Coupled with the scarcity of hydrogen peroxide, the designers adopted enhanced diesel-electric power. The hull was dramatically streamlined removing the exposed gun and any external feature that would heighten resistance as well as reducing the size and profile of the conning tower. Range and battery time were improved and underwater speed was raised to 18 knots, over 10 knots faster than before. Their design diving depth was improved dramatically. They arrived too late for war service.

The Type XXI featured an evolutionary design that combined several different strands of the U-Boat development program, most notably from the Walter U-boats, the Type XVII, which featured an unsuccessful yet revolutionary hydrogen peroxide air-independent propellant system.

The Germans built some novel submarine designs, including the Type XVII, which used hydrogen peroxide in a Walther turbine (named for its designer, Dr Hellmuth Walther) for propulsion. They also produced the Type XXII, which had a large battery and mechanical torpedo handling.

Following the end of World War II both the British and United States Navies acquired some Type 21s for evaluation. They were amazed at the advances in the German boat. The United States then played catch-up, and the Bureau of Ships adapted the extremely successful American fleet submarine of World War II for greater underwater speed by streamlining the hull and conning tower, removing all appendages, including the guns, and dramatically increasing the battery power. The submerged speed increased from 8.75 knots to 18.2 knots (Albacore). The Navy applied the results of its research with Albacore to an operational vessel called Barbel in the late 1950s. Save for nuclear power, these diesel electric vessels had almost everything that the latest knowledge of hydrodynamics and modern technology could provide. Barbel was the first operational submarine to have the Albacore hull design, a single most efficient propeller on the axis, an advanced ballast control panel and HY-80 hull steel. She displaced 2,145 tons with a length-to-draft ratio of 7.55 and could make 21 knots submerged.

POST WORLD WAR II

During the Cold War, the United States and the Soviet Union maintained large submarine fleets that engaged in cat-and-mouse games. This continues even today, on a much-reduced scale. The Soviet Union suffered the loss of at least four submarines during this period: K-129 was lost in 1968 (which the CIA attempted to retrieve from the ocean floor with theHoward Hughes-designed ship named Glomar Explorer), K-8 in 1970, K -219 in 1986 (subject of the film Hostile Waters), and Komsomolets (the only Mike class submarine) in 1989 (which held a depth record among the military submarines—1000 m, or 1300 m according to the article K-278). Many other Soviet subs, such as K-19 (first Soviet nuclear submarine, and first Soviet sub at North Pole) were badly damaged by fire or radiation leaks. The United States lost two nuclear submarines during this time: USS Thresher andScorpion. The Thresher was lost due to equipment failure, and the exact cause of the loss of the Scorpion is not known.

The sinking of PNS Ghazi in the Indo-Pakistani War of 1971 was the first submarine casualty in the South Asian region.

The United Kingdom employed nuclear-powered submarines against Argentina during the 1982 Falklands War. The sinking of the cruiser ARA General Belgrano by HMSConqueror was the first sinking by a nuclear-powered submarine in war. During this conflict, the conventional Argentinian submarine ARA Santa Fé was disabled by a Sea Skua missile, and the ARA San Luis claimed to have made unsuccessful attacks on the British fleet.

AIR INDEPENDENT PROPULSION SYSTEMS

Air-independent propulsion (AIP) is any marine propulsion technology that allows a non-nuclear submarine to operate without access to atmospheric oxygen (by surfacing or using a snorkel). AIP can augment or replace the diesel-electric propulsion system of non-nuclear vessels.

The United States Navy uses the hull classification symbol "SSP" to designate boats powered by AIP, while retaining "SSK" for classic diesel-electric attack submarines. Modern non-nuclear submarines are potentially stealthier than nuclear submarines; a nuclear ship's reactor must constantly pump coolant, generating some amount of detectable noise (acoustic signature). Non-nuclear submarines running on battery power or AIP, on the other hand, can be virtually be silent. While nuclear-powered designs still dominate in submergence times and deep-ocean performance, small, high-tech non-nuclear attack submarines are highly effective in coastal operations and pose a significant threat to less-stealthy and less-maneuverable nuclear submarines. AIP can be retrofitted into existing submarine hulls by inserting an additional hull section. AIP does not normally provide the endurance or power to replace atmospheric dependent propulsion, but allows longer submergence than a conventionally propelled submarine. A typical conventional power plant provides 3 megawatts maximum, and an AIP source around 10% of that. A nuclear submarine's propulsion plant is usually much greater than 20 megawatts. Advances in AIP notwithstanding, nuclear-powered submarines are expected to account for the largest share of the overall market.

After the end of the Second World War, all round effort was put in various countries to find air independent propulsion systems for submarines so that they could remain submerged for longer durations, thus avoiding the need to surface for running the diesel engines to charge the batteries and consequently get detected from the air and get bombarded. It was said that the next war would be won by the nation, possessing the submarines which would have no need to surface.

Air Independent Propulsion (AIP) technology became the holy grail of submarine design ever since German engineer Hellmuth Walter first began experimenting with his high purity hydrogen peroxide powered engine in the early 1930s. In the Year 1940, Helmuth Walter demonstrated a prototype for the first true submarine, a boat which in theory, could operate submerged for an indefinite period, unlimited by battery capacity or the need for atmospheric oxygen. V.80 was powered by the decomposition of highly concentrated (95 percent) hydrogen peroxide, H2O2, known as Perhydrol, by a potassium permanganate catalyst. In essence, when the chemical breaks down, it releases superheated steam to drive a turbine, along with oxygen to support conventional combustion or for respiration by the crew. Several experimental boats were produced, and one, U-1407, which had been scuttled at the end of World War II, was salvaged and re commissioned into the Royal Navy as HMS Meteorite. The British built two improved models in the late 1950s, HMS Explorer, and HMS Excalibur. Meteorite was not popular with its crews, who regarded it as dangerous and volatile; she was officially described as "75% safe". The reputations of Excalibur and Explorer were little better, the boats were nicknamed 'Excruciater' and 'Exploder'. The Soviet Union also experimented with the technology and one experimental boat was built which utilized hydrogen peroxide in a Walter engine. The United States also used hydrogen peroxide in an experimental midget submarine, X-1. It was originally powered by a hydrogen peroxide/diesel engine and battery system until an explosion of her hydrogen peroxide supply on 20 May 1957. X-1 was later converted to a diesel-electric.

The USSR, UK, and the US, the only countries known to be experimenting with the technology at that time, abandoned it when the latter developed a nuclear reactor small enough for submarine propulsion. Other nations, including Germany and Sweden, would later recommence AIP development.

Today, there are four basic technologies - closed cycle diesel, closed cycle steam turbines, Stirling engines and fuel cells - and each has its own peculiarities.

Closed Cycle Diesel Engine-This technology uses a submarine, diesel engine which can be operated conventionally on the surface, but which can also be provided with oxidant, usually stored as liquid oxygen, when submerged. Since the metal of an engine will burn in pure oxygen, the oxygen is usually diluted with recycled exhaust gas. Argon replaces exhaust gas when the engine is started. During World War II the Kriegsmarine experimented with such a system as an alternative to the Walter peroxide system, including a variant of the Type XXVIIB Seehund midget submarine, the "Klein U-boot". It was powered by a 95 hp Diesel engine of a type commonly used by the Kriegsmarine and which was available in large numbers, supplied with oxygen from a tank in the boat's keel holding 1,250 liters at 4 atmospheres (410 kPa). It was thought likely that the boat would have a maximum submerged speed of 12 knot (22 km/h; 14 mph) and a range of 70 miles (110 km), or 150 miles (240 km) at 7 knot (13 km/h).

The USSR built on the German work, developing the small 650 ton Quebec-class submarine of which thirty were built between 1953 and 1956. These had three diesel engines, two were conventional and one was closed cycle using liquid oxygen. In the Soviet system, called a "single propulsion system", oxygen was added after the exhaust gases had been filtered through a lime-based chemical absorbent. The submarine could also run its diesel, using a snorkel. The Quebec had three drive shafts, 900 hp diesel on the centre shaft and two 700 hp diesels on the outer shafts. In addition a 100 hp "creep" motor was coupled to the centre shaft. The boat could be run at slow speed using the centerline diesel only. Because liquid oxygen cannot be stored indefinitely, these boats could not operate far from a base. It was dangerous; at least seven submarines suffered explosions, and one of these, M-256, sank following an explosion and fire. They were sometimes nicknamed cigarette lighters. The last submarine using this technology was scrapped in the early 1970s.

Closed-cycle steam turbines-The French MESMA (Module d'Energie Sous-Marine Autonome) system is offered by French shipyard DCNS. MESMA is available for the Agosta 90B and Scorpène-class submarines. It is essentially a modified version of their nuclear propulsion system with heat generated by ethanol and oxygen. Specifically, a conventional steam turbine power plant is powered by steam generated from the combustion of ethanol (grain alcohol) and stored oxygen at a pressure of 60 atmospheres. This pressure-firing allows exhaust carbon dioxide to be expelled overboard at any depth without an exhaust compressor. As installed on the Scorpène, it requires adding a 8.3 meter (27 foot), 305 tonne hull section to the submarine, and results in a submarine, capable to operate for greater than 21 days underwater, depending on variables such as speed. An article in Undersea Warfare Magazine notes that: "although MESMA can provide higher output power than the other alternatives, its inherent efficiency is the lowest of the four AIP candidates, and its rate of oxygen consumption is correspondingly higher.

Stirling cycle engines- The Swedish shipbuilder Kockums constructed three Gotland-class submarines for the Swedish Navy that are fitted with an auxiliary Stirling engine that burns liquid oxygen and diesel fuel to drive 75 kilowatt electrical generators for either propulsion or charging batteries. The endurance of the 1,500-tonne boats is around 14 days at 5 knot (5.8 mph; 9.3 km/h). Kockums has also refurbished/upgraded the Swedish Västergötland class submarines with a Stirling AIP plug-in section. The Södermanland and Östergötland are in service in Sweden as the Södermanland class, and two others are in service in Singapore as the Archer class (Archer and Swordsman).

Kockums also delivered Stirling engines to Japan. New Japanese submarines will all be equipped with Stirling engines. The first submarine, Sōryū, was launched on 5 December 2007 and the remainder were delivered to the navy in March 2009. The new Swedish A26 submarine has the Stirling AIP system as its main energy source. The submerged endurance will be more than 18 days at 5 knots using AIP.

Fuel Cells-Siemens has developed a 30-50 kilowatt fuel cell unit. Nine of these units are incorporated into Howaldtswerke Deutsche Werft AG's 1,830 ton submarine U31, lead ship for the Type 212A class of the German

Navy. The other boats of this class and HDW's AIP equipped export submarines (Dolphin class submarine, Type 209 mod and Type 214) use two 120 kW modules, also from Siemens. After the success of Howaldtswerke, Deutsche Werft AG's in its export activities, several builders developed fuel-cell auxiliary units for submarines, but as of 2008 no other shipyard has a contract for a submarine so equipped.

The AIP implemented on the S-80 class of the Spanish Navy is based on a bioethanol-processor (provided by Hynergreen from Abengoa, SA) consisting of a reaction chamber and several intermediate Coprox reactors, that transform the BioEtOH into high purity hydrogen. The output feeds a series of fuel cells from UTC Power company, which also supplied fuel cells for the Space Shuttle. The reformator is fed with bioethanol as fuel, and oxygen (stored as a liquid in a high pressure cryogenic tank), generating hydrogen as a sub-product. The produced hydrogen and more oxygen are fed to the fuel cells.

Indian DRDO, the Defence Research and Development Organisation has developed an AIP system based on Phosphoric Acid Fuel Cell (PAFC) to power the last two Kalvari-class submarines which are based on the Scorpene-class submarine design.

As of 2015, some 8 nations are building AIP submarines with over 14 nations operating AIP based submarines. Australia wants a new Japanese propulsion system for its next generation of submarines ;(government officials with direct knowledge of the matter said), bolstering Tokyo's position as the likely builder of the multibillion-dollar fleet. Reuters reported in September that Australia was leaning towards buying 12 submarines based on Soryu-class vessels built by Mitsubishi Heavy Industries and Kawasaki Heavy Industries. The new submarines will replace six ageing Collins-class boats. In talks since then, Canberra has said it wants a lithium-ion battery propulsion system for the submarines, two Japanese officials and one Australian official told Reuters.

Japan is a leader in lithium battery technology and its next generation of Soryu submarines will be the world's first to be powered by such a propulsion system. Lithium-ion propulsion has not been used in part because of the difficulty in adapting the technology for use in confined vessels. (These batteries, in smaller version, have been successfully used for the cardiac pacemakers. Lithium iodine battery was invented and used by Wilson Greatbatch and his team in 1972 made the real impact to implantable cardiac pacemakers. This tiny battery, implanted in the cavity, near the collar bone of a patient, lasts for about 10 years and even today, is the power source for many manufacturers of cardiac pacemakers. The pacemaker unit delivers an electrical pulse with the proper intensity at the proper location to stimulate the heart at a desired rate. A cardiac pacemaker uses half of its battery power for cardiac stimulation and the other half for housekeeping tasks such as monitoring and data logging.)

The above vessels are expected to be tested and commissioned around the end of the decade. Experts say the technology will give submarines better underwater range and speed compared to other diesel-electric vessels that use air independent propulsion (AIP) undersea, as the system requires fuel to operate. The deal would also mark Japan's re-entry into the global arms market, just months after Prime Minister Shinzo Abe ended a ban on weapons exports as part of his efforts to steer Japan away from decades of pacifism. Prime Minister Tony Abbott is under mounting domestic pressure to build the submarines at home, especially since European contenders have emerged and said they would do the work in Australia.

Clearly, no AIP technology currently has all the answers, but the appeal is obvious. Allowing a submarine to remain submerged for long periods, without the need to surface, brings increased range and improved underwater endurance, without the high cost of nuclear propulsion, while retaining the advantages of conventional diesel electric power. In addition, the size of AIP boats allows them to operate effectively in littoral waters, where access is often a problem for their nuclear counterparts, making the technology particularly attractive to smaller regional navies and nations which lack the expertise, the budget, or the desire, to pursue the nuclear-powered route. The 'Global Submarine Market 2011-2021' concluded that mature AIP systems are increasingly seen as a 'must-have' capability, especially since the varying mission profiles of modern deployments now prioritize submarines as multirole platforms. AIP is, consequently, a developing field; the growing global interest in AIP is beginning to have a positive effect on technology development, but so far none of the technologies seem powerful enough to run a submarine on its own.

NUCLEAR PROPULSION IN US

After the discovery of nuclear fission in 1939, the original interest of the US Government in harnessing the power of atom was not to produce an atomic bomb but to devise nuclear propulsion for ships. Before the outbreak of the Second World War, the US Naval Research Laboratory was engaged in studying the idea of building a propulsion plant for ships. To build a light weight, small size nuclear plant for ships, the Laboratory developed the first method of separating the essential isotope of Uranium 235 from natural uranium. This was the thermal diffusion method which did not prove to be as practical as the gaseous diffusion later developed at Oak Ridge.

Once the decision was made to develop an atomic bomb, the idea of propulsion plant was shelved and the bomb project was given top priority; all the money and the skills of the country were devoted to making the atomic bomb. Nuclear propulsion for ships was put aside till the end of the war. After the war, when the subject of nuclear propulsion was again considered, there were people who thought that atomic reactors with large amount of shielding of steel and concrete could never be carried aboard ships. They had in mind very large reactors at Hanford, built for producing plutonium to build bombs during the war. Others thought that human beings could not live and work in close proximity of radiation. Fortunately there were others who thought that these problems could be overcome; the credit must go to them for the success of atomic submarines and even the first commercial nuclear powered ship 'Savannah' which was commissioned a few years after the first nuclear propelled submarine 'Nautilus'.

THE DELAY—Technical & Organisational

Towards the end of 1945, the idea of an atomic powered submarine was fascinating to Captain Rickover, who led the program and later became an admiral. Since the atomic reactor did not require oxygen, for the first time, the navy could have a true submarine which could stay underwater indefinitely without contact with the surface, the range limited to the stamina of the crew. Advantages were tremendous; it could sneak at an enemy surface ship at full speed, hit the enemy ship and turn and escape submerged, without the enemy knowing. With such high underwater speed, it could outrun the enemy surface ships, cross oceans completely submerged, thereby reducing the number of submarines required in a fleet. The atomic submarine would fulfill all dreams of Jules Verne, who created the famous mythical submarine Nautilus in 'Twenty thousand leagues under the sea' Captain Rickover concluded that the atomic powered submarine was the first order of business for the United States Navy. If a reactor could be built to fit in a small submarine, it was sure to be fitted in any larger ship.

Many outstanding engineering achievements, in the world, were initially delayed, not only due to lack of financing, but also due to lack of confidence, faith and encouragement; submarine propulsion is an example. Some of the admirals and scientists were in favor and others against this program; they used to have discussions and arguments regarding the urgent need for nuclear propulsion for the submarines; though the actual war had ended but the cold war had set in. Many naval officers were opposed on the grounds that it had no logical place in the Navy's arsenal of weapons. In 1947, the General Electric shifted its emphasis from the naval reactor to 'breeder' reactor. Rickover's first concern was to allay fears of those who saw the submarine project as a threat to the production of nuclear weapons for the country's defense and others who saw it as a blow to the promise of the civilian nuclear power programme. He could convince the persons concerned that the submarine project like hydrogen bomb would become key to the nation's defense; civilian nuclear power programme would become easier once the nuclear propulsion is developed.

When Captain Rickover was made the 'special assistant for nuclear matters' in the 'Bureau of Ships', it became apparent to him that someone has to shake up the 'Navy' and the 'Atomic Energy Commission' at the highest level for the project to start and that he would have to do it alone. The strategy was to get the Secretary of Navy to declare the atomic powered submarine as a military necessity, so that the Atomic Energy Commission would be forced to divert some efforts from bombs to reactors. By this time, there was no effort on designing a power reactor for electricity generation. Though a number of reactors had been built in the US at Hanford and Oak Ridge, they were huge; the power generated per kilogram of weight of the reactor system was too small and thus was impractical. Engineering miracle was required whereby these massive reactors could be scaled down and compressed into a small space.

Captain Rickover had by then concluded that the reactor will have to be thermal/slow neutron type and it should have water as coolant in the reactor. In all of the classical Physics and engineering, no material was more commonly used than water. Its abundance, properties and man's experience in using it, made water as the most attractive material for heat transfer medium in a reactor. Yet while reviewing the existing data on water technology, Rickover and his colleagues were surprised to discover how little was known about the properties of water and its effect on materials. There was little understanding of how metals or oxygen got dissolved or suspended in boiling water in conventional steam plants. Corrosion of stainless steel systems by water were even less known. Changes in temperature, flow rate, chemical composition of water could create deposits, and would carry radioactivity to the external portions of the plant and adversely affect the heat transfer; in addition, sticking and galling of parts of the mechanisms were observed. A number of laboratory studies were carried out on corrosion and wear in water systems.

He also discussed the tremendous additional problems of squeezing down the reactor so that it might fit into a submarine. He arranged a meeting between the AEC officials and the 'Chief of Bureau of ships' to discuss in broad terms, the inherent problems in the nuclear powered ship program. He then took the help of Admiral Nimitz, Chief of naval operations to write a letter to the Secretary of the Navy. The letter mentioned that the great strategic and tactical advantage of nuclear powered submarine should be brought to the attention of the Secretary of Defense and the Research and Development Board of the Department of Defense; the Research & Development Board, Bureau of ships and the Atomic Energy Commission should, together work out the pending action, mutually agreeable, for design, development and construction of the nuclear propulsion plant for submarines. After obtaining the high power backing from the Navy and Defense Research & Development Board, he decided to tackle the Atomic Energy Commission, by requesting the Chief of Bureau of Ships to write to the Atomic Energy Commission. The letter stated that the responsibility for atomic power was, by law, vested in the Atomic Energy Commission; the responsibility of the navy is to have best possible naval vessels; to implement the national policy, it was mandatory that no time be lost in proceeding with the design of submarine nuclear power plant. The letter added that the problems to be solved are so intimately connected with both atomic energy and the navy that neither can make separate engineering decisions and the program must be co-ordinated in the closest possible manner between them. The basis of the plan suggested was that the authority for this work be delegated to a single organization working for both, the Bureau of ships and the Atomic Energy Commission and the facilities of the Atomic Energy Commission, the Bureau of ships and the office of the Naval research be used to the fullest extent. This was the beginning of 1948.

The Atomic Energy Commission was still not fully convinced. During the 'undersea warfare symposium', where more than 700 people, including leaders of US scientific world had gathered, papers were read and open criticism was done. The Atomic Energy Commission was advised to develop the program that can enlist all varieties of industrial and academic institutions and encourage them to operate under separate directions and with their own separate slants; the Atomic Energy Commission should not hold all the reins tightly for taking all decisions. All industrial and laboratory agencies that can contribute to the solution of this project must be identified and used. At this stage the Westinghouse Electric Corporation was brought on board; Westinghouse was required to set up a laboratory with a requisite scientific staff to carry out experiments on a heat exchanger system for an atomic power reactor, similar to the General Electric program, but the difference was that Westinghouse would be working with pressurized water as the coolant instead of low pressure sodium, being used by General Electric in the heat exchanger system. Sodium as compared to water was a mystic heat transfer medium; not much information was available. Office of the naval research and the Atomic Energy Commission cooperated in preparing a hand book of all available information on using liquid metals for heat transfer. This handbook, The Liquid Metals Handbook, first published in 1950 became the first of many handbooks on Reactor Engineering which later appeared over the next ten years. This literature became the essential part of the technical foundation not only for the nuclear propulsion program, but also for reactor development in general; this literature is used even today and is the back bone of the engineering design of 'fast reactors'.

Conducting these experiments was not enough; they had to work on the most difficult task, the reactor itself. This was step in right direction; two of the largest and most capable electric firms were in competition each other in atomic power business; if simulated, they might soon plunge into large scale undertakings of constructing reactors.

Equally important, like the heat transfer medium in a practical nuclear propulsion plant, was the shielding which would protect the personnel from the extraordinary amounts of radioactivity generated within the reactor plant. During the year at Oakridge, Rickover realized that the massive shielding used in the land based Plutonium production reactors at Hanford provided very little practical experience in designing the effective shielding for the submarine plant. He requested two scientists Libbey and Blizard to compile a technical summary of information on shielding. The report explained the different types of radiation and the possibility of constructing different shields from different combinations of materials. Question of shielding was so vital that Rickover insisted on examining all the fundamental assumptions involved, including the accepted standards for radiation protection. He even accepted the possibility that a very low level of radiation might later be found to have some effects on man. He even consulted world famous geneticists to discuss radiation effects. Thus the firm understanding of the subject helped to take a practical and effective approach to shielding design. A series of experiments were performed in research reactors, to test the performance of various shielding materials; these experimental results were then translated into the shielding design. The shielding group continued to compile the basic data which later appeared in the Reactor Shielding Design Handbook which is now referred for any design of a system, involving nuclear applications, including radiography, nuclear medicine or radiotherapy.

There was a challenge to the metallurgists; this was to find a metal which could withstand high stresses while operating at high temperatures and also not absorb neutrons; this was a strange new metal called zirconium and the large scale production had to be started. Zirconium's low affinity for neutrons, promised more efficient use of uranium, which was considered the scarce resource. In addition to Zirconium, the project also required substantial quantity of Beryllium and later Hafnium. In contradiction to zirconium, hafnium, unlike zirconium, absorbed neutrons and was capable of withstanding high temperatures; this was also required in reasonably large quantities. All these materials were relatively unknown in American Industry and their use would require extensive study of their physical, chemical, metallurgical and nuclear properties. For each metal, the research included a study of the ore bearing these materials, methods of extraction, the process of reducing the material to metal and special techniques for fabricating, treating and testing the metal. It was also necessary to establish commercial facilities to produce these materials or to develop new processes suitable for large scale production. A number of laboratories and research institutions involved in the fundamental studies needed to be coordinated. The technical data for both, water cooled thermal reactor and sodium cooled intermediate reactor, was produced. A number of handbooks were later compiled and produced; some of these are 'A bibliography of Reactor Computer codes', 'The Metallurgy of Zirconium', 'The Metal Beryllium', 'Corrosion and Wear Handbook', 'The Metallurgy of Hafnium' and 3 volumes of 'Physics Handbook'. These handbooks became the building blocks of the new technology.

With Navy and the Atomic Energy Commission backing the program along with two large industrial firms at work, Captain Rickover turned his full attention to another major problem, the creation of a joint navy atomic energy task force to supervise and co-ordinate the growing list of activities. In August 1948, a new Nuclear Power Division of the Bureau of Ships was regrouped under Captain Rickover. At the same time, although preoccupied with the primary role of making better atomic bombs, the Atomic Energy Commission in September 1948, decided to establish a full size 'Division of Reactor Development' though it was actually formed in February 1949. Five month delay was primarily caused because no one could be persuaded to come forward and take over the new division for designing different kinds of power reactors; finally Dr Hafstad was persuaded to take over. Captain Rickover went to see him and gave him one of the most elaborate briefings on atomic submarine. He asked Dr Hafstad's permission to transfer the Naval Group bodily into the Atomic Energy Commission so that it could work directly on the nuclear ship program. Impressed with the aggressiveness and persistence and after getting strong recommendations about Rickover from others, Dr Hastad agreed that the new division of Reactor Development should include the Naval Reactor Branch headed by

Captain Rickover. Four studies would be undertaken by this Division of Reactor Development; the first would be a submarine thermal reactor using water as a coolant, the second would be General Electric's breeder reactor, the third would be a small 'material testing reactor' to be used for testing the metals for corrosion and other characteristics, the fourth would be the 'experimental power breeder'.

Thus Captain Rickover became the 'Chief' of the naval reactor branch of Atomic Energy; the position became very valuable for both organizations. It took a lot of discussion to break down the barriers of hostility, trust and misunderstanding which had grown during the months of pressure and frustration on the project. The members were able to talk to each other instead of commenting on each other; they started discussing the project objectively. Red tape and frustrating delays were cut to the minimum and most of the effort was spent on solving the technical problems. Westinghouse got fully involved as they were made to realize that a profitable future lay ahead for the company if it could develop atomic power plants for industrial use and the logical way to embark on such a program was to first build reactors for naval propulsion, thus gaining valuable experience. Later the shift to industrial reactors would be easy. This meant that some fissionable material would be diverted from the atomic bomb production and allotted to the power reactor program. The contract between the Atomic Energy Commission and the Westinghouse was signed; Atomic Energy agreed to furnish funds for the construction of an atomic powered laboratory in conjunction with the scientists of Argonne National Laboratory of the Atomic Energy. This was built by Westinghouse at a site outside Pittsburg; the required physicists and the engineers were located and assembled, to design and build thermal water cooled submarine reactor.

The design of the 'Reactor Core' consisting of fuel elements, control and shut off rods mechanisms, housed in the reactor pressure vessel is probably the most difficult task for the reactor engineers. Evolving a practical design of fuel elements proved to be an especially difficult task. Not until March 1950, did Argonne and Bettis decide that it would be feasible to assemble a fuel element consisting of uranium zirconium alloy clad with zirconium. Once the decision was made, the laboratories struggled together to arrive at a workable process to fabricate hundreds of fuel elements required for the reactor. Even the processing of metal proved tricky; it would pick up contaminants during processing and fabrication and would destroy its corrosion resistant properties. Similar precautions were necessary in forming the uranium alloy and then bonding the material to zirconium cladding. The uncertainties surrounding the development of control rod mechanisms were even more complex. The control rods containing neutron absorbing material would move in grooves in the fuel elements. There was a straightforward problem of choosing the best material; it was decided to use an alloy of silver and neutron absorbing material cadmium which would be bonded to stainless steel strips by hot rolling. Later cadmium was replaced by hafnium available in abundance by the processing of zirconium. The design of drive mechanisms for the control rods was complicated as it had to meet conflicting requirements. On one hand it was advantageous to control each rod individually in order to achieve optimum power distribution in the core for throughout its life; on the other hand, the rods had to operate within the reactor pressure vessel. Both the entire drive mechanisms and the motors had to operate inside the pressure vessel or the leak tight seals had to be developed to connect the external portions of the mechanisms and motors to the control rods moving inside the vessel. So the number of drive shafts was reduced by 'ganging' the control rods and connecting these with rack and pinion arrangements inside the pressure vessel. Later review indicated that there was no valid reason for continuing work on the 'ganging' system; the complicated mechanical arrangement would never be reliable nor it could be maintained once the reactor became 'critical'. Finally, individual drives with external motors were used and the seals were developed for the shafts against the leakage of water and radioactivity.

Organisation Structure

The layman's common impression of nuclear sciences was that they involve extremely complex and esoteric conceptions that were far beyond the understanding of ordinary men. In some areas of nuclear physics this impression was correct, but in other areas the real ability in engineering was required.

To review the progress 'Nuclear Power Division' set up Code 390, Rickover was the most important member. After complete agreement, the priorities and the goals were set; only the research for production plants at Hanford would take precedence over the submarine project. Manpower was shifted from the breeder project to submarine project; the aim would be to build a land based sodium cooled prototype as soon as possible. Within the General Electric organization itself, Rickover had some success in establishing a distinctive structure for the project and clear lines of authority. The basis of Rickover's authority was his dual role which tied him to both the commission and the navy; he was able to turn this to his advantage. Instead of double infringement on his authority, the dual organization became a vehicle for unusual independence. Sometimes he would act a naval officer and sometimes a commission official. The diversity prevented Rickover from following any uniform or fixed pattern of organization. He never established a rigid system; he passionately believed that the success in building the project lay in flexibility. He would not be committed to a fixed structure; every new situation involved a unique combination of personalities, talents and technical considerations. Rickover accepted the decentralization of design and procurement functions as essential in modern technology but he insisted on retaining tight controls on field activities. He was very careful to select those who would demonstrate some practical knowledge and skill in engineering. There had to be more stress on reactor technology and less on nuclear physics. He assured that the decisions were made on a sound technical basis. It was all too easy, especially in a military organization, for juniors to defer to the seniors even when they knew that the decision was ill founded. Rickover had earlier suffered superiors who had made technical decisions on matters which they did not understand and then arranged to present their opinion in such a way that no one would dare to contradict them. His refusal to accept the decisions when he was convinced that they were wrong had made him unpopular earlier.

Code 390 as it emerged in the early 1950, reflected Rickover's personal experience and his philosophy of technical judgment. The task in his view was one for engineers rather than administrators; it was for men who could understand the intricacies of design and manufacturing, who take the initiative in engineering and direct the work of contractors. Thus the form of the organization at any given time rested on the status of development at that particular time as it did on the qualifications of the engineers assigned.

As a professional naval officer who had spent large part of his career in Washington, Rickover was familiar with the ways of the navy and the bureau of ships. Compared with the Atomic Energy Commission, the navy seemed old fashioned, un enlightened and tradition bound bureaucracy whose organization and methods were not equal to the task of exploiting the advantages of modern technology. Rickover hoped that he would use the project to convince the Navy to accept some of the methods and approaches he was using. During the next few years, he never ceased to challenge the old ideas and the prejudices or to propose new approaches and methods.

Many aspects of construction encountered were similar to any other project which had to meet the strict time schedule, but one aspect which posed difficulty was uncommon to most construction projects. This was the exceptional cleanliness required of all components to be installed in the plant. Fabricators had to follow special procedures during manufacture and inspection to ensure that no foreign matter was introduced. Special wrapping and tagging regulations, unfamiliar to most industries, were introduced. The teams were sent to manufacturers' plants to inspect cleanliness procedures.

Building of the Land Based Unit and Nautilus

Submarine reactors need to be robust and resilient enough to withstand the rigors of decades at sea. The physical stresses during onboard operations, along with the potential for rapidly changing power demands, and the harsh conditions within the reactor plant itself, combine to produce a uniquely challenging set of circumstances for nuclear propulsion. Scaling down and compacting the reactor was a challenge to the engineers. The required number of good engineers had to be found and trained. With the firm backing of navy and Atomic Energy, the educational staff at the Massachusetts Institute of technology arranged a special one year course in nuclear engineering; the graduates were integrated into the navy atomic energy nuclear power program. There was no strict caste system between the civilians and naval officers; working atmosphere befitting their technical competence was established. The Westinghouse was explained that the atomic powered submarine is 95 % engineering and 5% physics; a series of 'indoctrination lectures' were designed for the Westinghouse company after getting the security clearance for the personnel. New York Times carried a news item—"Executives of the Westinghouse Electric Corporation are now attending an 'atom school', 17 weekly lectures on nuclear engineering, in Pittsburg. Sponsors of the school are Atomic Energy Commission and the Westinghouse Atomic Power Division, which now is engaged in procuring a nuclear power plant to drive a navy ship. The 45 students include a number of elected officers of the company and top executives from Pittsburg headquarters and East Pittsburg works; the professors are naval officers associated with AEC's Reactor Development Program".

In the guarded buildings of Bureau of Ships in Washington, the various parts and portions of the reactor started taking shape, secretly, on the drawing boards. Though the US had been building the submersible ships, which they called submarines, none of those hulls, could be really called a true submarine, designed primarily for prolonged submerged operations. A very big advantage, strange it may sound to non-technical persons, is that per given unit of power, the ships at high speeds can move faster underwater than they can on the surface. The biggest problem was to fit the bulky atomic reactor inside the hull; some felt that the reactor should be partly outside the hull away from the ship, like aircraft engine which is in the wings and others argued that the reactor should be inside the hull in the designated 'reactor compartment' thus containing all the radioactivity inside the compartment. Finally, Westinghouse solved the problem of accommodating the reactor in the hull at the same time solving and simplifying the maintenance problems when it was discovered that the coolant water lost its radioactivity rather quickly once the reactor is shut down thus permitting rapid access to the machinery. When a brand new type of propulsion plant is built for a ship, aircraft or any other vehicle, it is customary to first build an experimental version on land to ensure that it performs properly. The naval group, Westinghouse and Argonne made the plans for the land based plant; the suggestion was to construct it in the traditional manner of building various parts and subsystems spread out in a building and after making sure that all subsystems worked well, these could be later rebuilt and fitted in the submarine. Captain Rickover ruled out this proposal and insisted that the land based would be built inside the submarine hull on a dry land; this looked extraordinary. Captain had his reason; if Westinghouse painstakingly built the land based, checked it thoroughly and then redesign to fit in the submarine, the two stages would take years to complete. The captain's idea was to build two plants simultaneously, one to be built as the land based prototype and the second would go into the submarine. The schedule would be worked out in such a way that the sea going reactor would lag slightly behind the land based reactor so that the changes and improvements found necessary could be quickly and timely implemented on the sea going version. Rickover's approach could be described as 'concurrent' as opposed to sequential development. There was a lot of discussion on the location; the site had to be away from the populated areas in case there would be an accident, at the same time it should not lead to the transportation problems for men and material; even building the plant on a large ship at sea, was also considered which was negated on the grounds of the risk of losing expensive and scarce material at sea. Finally the site chosen was Arco Desert in Idaho.

There were two submarine yards on the east coast, which had been manufacturing submarines on a large scale during the Second World War. One was the Electric Boat Company and the other was the navy's own submarine building yard at Portsmouth, New Hampshire. Traditionally, Westinghouse used to manufacture the electrical equipment for the submarines, built by the Navy at Portsmouth and General Electric used to manufacture the electrical equipment for the submarines made by the Electric Boat Company. Captain Rickover hoped to continue the similar relationship for the atomic powered submarine program. He wanted to have two pairs for the submarine program; Westinghouse & Portsmouth to build one atomic submarine and General Electric & Electric Boat Company to build another atomic submarine. But he could not succeed as the Portsmouth yard could not take up the job, mainly due to lack of funds and insufficient staff. He could finally convince the Electric Boat Company to be a subcontractor to the Westinghouse Electric Corporation in designing and building the desert bound land based prototype submarine and its machinery. General Electric had been experimenting with their 'breeder' reactor, which would create fuel as it consumed; Captain Rickover convinced the GE that they should pay attention to the naval reactor instead of the 'breeder' reactor program. It became apparent to GE that if they were to prosper in the nuclear power business, like its rival Westinghouse, they must participate in the atomic submarine program. Finally, GE reoriented the program with emphasis on the submarine intermediate reactor and the power breeder reactor. Once again, the Electric Boat Company readily agreed to assist the GE in the ship building aspects of the project as they were doing for the Westinghouse. To avoid confusion, labels were put on both the projects; the Westinghouse project was named STR short for 'submarine thermal reactor' and the GE project as SIR for 'submarine intermediate reactor'. The land based reactor and the sea going version of the Westinghouse STR, were named Mark One and Mark Two respectively, while the GE's SIR were named Mark A and Mark B for the land based and sea going version respectively. Thus, two of the largest and most competent industrial firms in the US took on the complex task of building not one but two atomic powered submarines. Two teams, for supervising the program and the working schedule were established, one for STR and the other for SIR.

Mark A design had some disadvantages; it would require more fissionable material than the Mark 1 for comparable power capacity, but more drawbacks were linked to the use of sodium as a coolant. Neutrons within the reactor would transmute some of the coolant into sodium 24 an isotope which has a half-life of 15 hours and emits gamma radiation of high energy. As a result every equipment and pipe in the primary system would need to be shielded and unlike Mark 1 plant; the reactor compartment could not be entered for maintenance until many hours after reactor shutdown. Sodium is also solid at room temperature that means it would freeze in pipes if not heated continuously when the reactor is not operating. In addition, sodium reacted violently with water. Although the laboratories had learnt a lot about minimizing the dangers of sodium, a leak in the system making sodium react with water could be disastrous. In spite of all these disadvantages, Mark A was too promising to be overlooked. The final decision was as follows.

Mark 1—Contractor Westinghouse, land based unit at Idaho, moderator and coolant water, fuel Uranium 235. Mark 2—To be fitted in Nautilus, to be built by Electric Boat Division, Groton.

Mark A---Contractor General Electric, land based unit in West Milton N Y, moderator Beryllium, coolant sodium and fuel Uranium 235. Mark B—To be fitted in Seawolf to be built by Electric Boat Division, Groton.

Out in the desert, near Arco, Idaho, a big and high concrete building was designed to shelter the Mark One, the Westinghouse land based prototype; for the first time a man was attempting to build a submarine in a desert. More than 2500 miles away, in Groton Connecticut, workers at the Electric Boat Company roped off an area nicknamed it 'Siberia' and began constructing a wooden 'mock up' , an exact replica in wood of the Mark One hull and the reactor plant. Wooden mock-ups have been used for years in construction of submarines to try the preliminary fitting of parts and pieces of machinery. By 1951, Mark One was under construction at Arco and Mark Two was about to be started at Westinghouse. Electric Boat Company started the plans for construction of the submarine itself. Nuclear system and the steam system raised a number of problems because there was not enough space inside the hull, for the required manpower to be deployed in every shift, for quickly fitting out the equipment. Another essential problem which was solved was the removal of CO2 from the air and the introduction of fresh Oxygen made from the distilled sea water. To confirm the efficacy of man-made air, a live test was performed on a number of sailors who breathed this air for six weeks. Westinghouse and Electric Boat Company conferred almost daily basis on some new problem before starting the construction. Finally the keel of the 'Nautilus' was laid on June 14, 1952 in Groton, Connecticut. In May 1953, the land based prototype was started in the desert Arco and it performed much better than the expectations and started the 'nonstop' long runs. The sparkling performance of Mark one suggested that the dream of a fleet of nuclear powered ships would be a reality.

The work on 'Nautilus' progressed very rapidly. The submarine was launched on January 21, 1954. The people, thousands of men, women and children, poured into the 'Electric Boat Yards'; it was like pushing one's way into a football stadium and inside hawkers were passing out the gaudy , gold colored program. Electric Boat had done a wonderful job of converting a colorless building-yard and overhead crane into a facsimile of a football stadium with colorful flags hanging around the place. A Navy commander looked up and saw Rickover sitting in the first row. Pointing to the other Navy brass sitting behind Rickover he exclaimed 'look at all those bastards there celebrating; only years ago they were trying to kill both, the old man and the project' Anyway it was a big day for the submariners. The Nautilus could be used for many other purposes besides her primary role of gaining and maintaining control of the seas. She could be assigned to special missions like espionage work, landing spies on enemy beaches, reconnaissance or for landing commando teams. The traditional launching ceremony broke the steady pace of around the clock activity on board. The discovery of a faulty steam pipe threatened the effort to complete the ship by January 1955. After replacing the pipe and with a last minute burst of effort, the team brought the Mark two to criticality on December 30. On the second day of the New Year, the reactor supplied steam for the ship's electrical system. Later that day the steam was fed into the turbines and the propellers turned over. On January 3, 1955 the Mark two reached full power.

Land based Prototype, May 1953

Nautilus, January 1954

On January 17, 1955, as the Nautilus slipped down the Thames, a signalman on the submarine blinked to the escort tug 'Skylark' "UNDER WAY ON NUCLEAR POWER". This and the subsequent trials exceeded expectations. During the first trial, Nautilus was confined to surface runs; submerged tests a few days later were more comfortable; new records were set for the submarine sea trials.

Seawolf

The GE's propulsion plants with 'submarine intermediate reactor' named earlier as Mark A and Mark B for the land based and sea going version also got commissioned soon thereafter. The Submarine Intermediate Reactor (SIR) nuclear plant was designed by General Electric's Knolls Atomic Power Laboratory and prototyped in West Milton, New York. The prototype plant was eventually designated S1G and Seawolf 's plant as S2G. Seawolf's keel was laid down on 7 September 1953 by the Electric Boat division of General Dynamics Corporation in Groton, Connecticut. She was launched on 21 July 1955, and commissioned on 30 March 1957. Seawolf was the same basic design as her predecessor USS Nautilus , but her propulsion system was more technologically advanced. It was carrying liquid sodium and epithermal, superheated, more powerful reactor and steam power plant, while Nautilus carried the alternative light water reactor and saturated steam plant. In Seawolf, the size of the machinery in the engineering spaces was reduced by about 40%. Her liquid-sodium cooled epithermal reactor was more thermally efficient than a light water-cooled system, quieter, and presumably better system. But she posed, presumably and arguably, several safety hazards for the ship and crew. Primary system pressure was 15 psi (100 kPa) and the only moving part in the primary system was the liquid sodium which was magnetically pumped by electromagnets external to primary piping. The phrase "Blue Haze" was often associated with the boat, which was Cherenkov radiation, visible on a dark night, in the sea water surrounding the hull, outboard of the reactor compartment. This was during the decay of radioactive 24Na (sodium-24) in the primary system over essentially the first half-life (15 hrs). There was only one coolant leak ever noted, and that was during fitting out in the yards. However the super heaters suffered from tube sheet welding cracks which allowed high pressure steam to leak into the low pressure primary system and react with the sodium to form sodium hydroxide and H2. Sodium also has a small fission capture cross section which formed 3H as a free gas in the primary system. This complicated system operation since 3H is radioactive, would mix with the H2 from Na-H2O reaction and had to be contained.

The Atomic Energy Commission historians' account of the sodium-cooled reactor experience was as follows;

Although makeshift repairs permitted the Seawolf to complete her initial sea trials on reduced power in February 1957, Rickover had already decided to abandon the sodium-cooled reactor. Early in November 1956, he informed the Commission that he would take steps toward replacing the reactor in Seawolf with a water-cooled plant similar to that in the Nautilus. The leaks in the Seawolf steam plant were an important factor in the decision but even more persuasive were the inherent limitations in sodium-cooled systems. In Rickover's words they were "expensive to build, complex to operate, susceptible to prolonged shutdown as a result of even minor malfunctions, and difficult and time-consuming to repair. The S2G reactor was replaced with a pressurized water reactor similar to Nautilus and designated S2W, the replacement process lasting from 12 December 1958 to 30 September 1960.

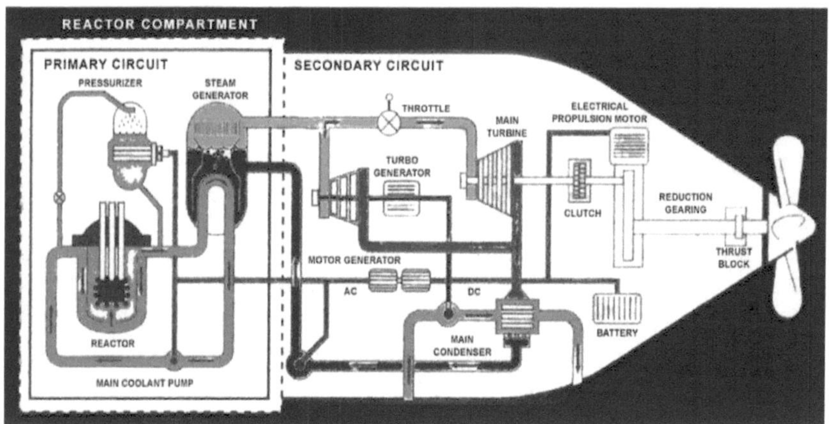

Typical 'flow sheet' of a 'nuclear propulsion plant'

Cold War— Strategic deterrence

Another revolution in submarine warfare came with USS George Washington (SSBN-598). Nuclear-powered, like Nautilus, George Washington added strategic ballistic missiles reaching the nuclear triad. Earlier submarines had carried strategic missiles, but the boats had been diesel powered, and the missiles required the boat to surface in order to fire. The missiles were also cruise missiles, which were vulnerable to the defenses of the day in a way that ballistic missiles were not. George Washington's missiles could be fired while the boat was submerged, meaning that it was far less likely to be detected before firing. The nuclear power of the boat also meant that, like Nautilus, George Washington's patrol length was limited only by the amount of food the boat could carry. Ballistic missile submarines, carrying Polaris missiles, eventually superseded all other strategic nuclear systems in the Navy. Deterrent patrols continue to this day, although now with Ohio-class submarines. Next to the aircraft carrier, the submarine remains a powerfully effective war time instrument capable of attacking naval assets and land-based targets from far-off locations

During the Cold War, the Soviet submarine program was a force to be reckoned with. The U.S.S.R.'s underwater killing machines captured the imagination of Westerners and Soviet citizens alike. Tom Clancy's 1984 novel (adapted as a film the following year) The Hunt for Red October depicted a daring attempt by the crew of a fictitious Soviet Typhoon Class ballistic missile submarine to defect to the United States At the height of the Cold War, approximately five to ten nuclear submarines were being commissioned from each of the four Soviet submarine yards. From the late 1950s through the end of 1997, the Soviet Union, and later Russia, built a total of 245 nuclear submarines, more than all other nations combined. Today, six countries deploy some form of nuclear-powered strategic submarines: the United States, Russia, France, the United Kingdom, China, and India are operating nuclear submarines. Several other countries, including Argentina and Brazil have ongoing projects in different phases to build nuclear-powered submarines.

Post Cold War

After the collapse of the Soviet Union, the Russian submarine program fell into decline along with many other branches of the Russian military. In the past decade, however, Russian officials have undertaken efforts to modernize their armed forces. From upgrading Cold War models to meet present-day challenges, to designing completely new platforms like the Borei and YasenClass submarines, Russia is clearly determined to renew the status and capabilities of its underwater fleet.

The Soviet and Russian nuclear submarines are divided into four generations according to their construction and combat capabilities. Russia's fifth-generation strategic and attack submarines will most likely be non-nuclear-powered, more compact and less "visible," a senior designer at the Rubin design bureau said 11 November 2013 Monday in an exclusive interview with RIA Novosti. Large nuclear-powered vessels, including Russia's Typhoon-class strategic boats, have so far dominated past and current trends in combat submarine construction. "I believe future submarines will be smaller, because of the use of more advanced technologies as well as the pursuit of more cost-effective production," Sergei Sukhanov said. "The fifth-generation boat will also be less 'visible' compared with existing submarines. They could also feature a new power plant, including fully electric," Sukhanov said, adding that changes could affect other sub-systems of future submarines. The designer said the most likely substitution for a nuclear reactor on strategic and attack submarines would be an air-independent propulsion plant (AIPP), which would make them stealthier than nuclear-powered boats. The AIPP allows a non-nuclear submarine to operate without the need to access atmospheric oxygen.

The Largest Submarine in the US Navy, USS Pennsylvania is the ballistic missile submarine which has been in commission since 1989. It has submerged displacement of >18,000 tons, it has a length of 170 meters, beam of 13 meters, draft of 12 meters; speed of > 25 knots and depth > 240 meters.

Supercarrier is an unofficial descriptive term for the largest type of aircraft carrier, typically those displacing over 70,000 tons . Supercarriers are the largest warships ever built, larger than the largest battleship class laid down by any country. The United States Navy has ten active nuclear propelled, supercarriers, each more than 100,000 tons displacement and > 330 meters long.

NUCLEAR PROPULSION IN SOVIET UNION

The important questions for the West in early fifties were about the capability of Soviet Union. Is the Soviet Union technically capable of producing an atomic powered submarine? If so, has the Soviet Navy the background and experience to make good use of it? Is the Soviet Union capable of waging effective anti-submarine warfare against atomic powered submarines of the West such as 'Nautilus?' The first question was partially answered a few months after the launch of 'Nautilus' when it was announced that the Soviets had built and successfully operated 'the world's first peace time atomic power reactor.' The significance of this announcement which was initially missed was the fact that, after successfully developing the atomic powered reactor, the Soviets had crossed the most difficult hurdle in building an atomic powered submarine.

Obninsk Nuclear Power Station was built in the "Science City" of Obninsk, Kaluga Oblast, about 110 km southwest of Moscow. It was the first grid-connected nuclear power station in the world, i.e. the first nuclear reactor that produced commercial electricity, albeit at small scale. It was located at the Institute of Physics and Power Engineering. The plant is also known as APS-1 Obninsk (Atomic Power Station 1 Obninsk). It remained in operation between 1954 and 2002, although its production of electricity for the grid ceased in 1959. Thereafter it functioned as a research and isotope production plant only. According to Lev Kotchetkov, who was there at the time: "Although utilisation of generated heat was going on, and production of isotopes was even enhanced, the main task was to carry out experimental studies on 17 test loops installed in the reactor. The technology perfected in the Obninsk pilot plant was later employed on a much larger scale in the RBMK reactors. The single reactor unit at the plant had a total electrical capacity of 6 MW and a net capacity of around 5 MWe. Thermal output was 30 MW. It was a prototype design using a graphite moderator and water coolant. The Obninsk reactor used 5% enriched Uranium; this percentage would be lowered for subsequent reactors. Construction started on 1 January 1951. First Criticality was achieved on 6 May 1954, and the first grid connection was made on 27 June 1954. For around 4 years, Obninsk remained the only nuclear power reactor in the Soviet Union; the power plant remained active until April 29, 2002 when it was finally shut down. According to Kotchetkov, in its 48 years of operation there were no significant incidents resulting in personnel overdose or mortality, or radioactive release to the environment exceeding permissible limits. The next Soviet nuclear power plant to be connected to their grid was Beloyarsk Unit 1 in 1964 with a capacity of 100 MWe.

Soviet Navy and Submarine History in Soviet Union

The Soviet Navy was based on a republican naval force formed from the remnants of the Imperial Russian Navy, which had been almost completely destroyed in the Revolution of 1917, the Russian Civil War, and the Kronstadt rebellion. During the revolution, sailors deserted their ships at will and generally neglected their duties. The officers were dispersed (some were killed by the Red Terror, some joined the "White" (anti-communist) armies, and others simply resigned) and most of the sailors left their ships. Work stopped in the shipyards, where uncompleted ships deteriorated rapidly. At the end of April 1918, German troops entered Crimea and started to advance towards the Sevastopol naval base. The more effective ships were moved from Sevastopol to Novorossiysk where, after an ultimatum from Germany, they were scuttled by Lenin's order. The ships remaining in Sevastopol were captured by the Germans and then, after November 1918, by the British

The first ship of the revolutionary navy could be considered the rebellious Imperial Russian cruiser Aurora, whose crew joined the Bolsheviks. Sailors of Baltic fleet supplied the fighting force of the Bolsheviks during the October Revolution. Some imperial vessels continued to serve after the revolution, albeit with different names. As the country's attentions were largely directed internally, the Navy did not have much funding or training. An indicator of its reputation was that the Soviets were not invited to participate in negotiations for Washington Naval Treaty, which limited the and capabilities of the most powerful navies During the 1930s, as the industrialization of the Soviet Union proceeded, plans were made to expand the Soviet Navy into one of the most powerful in the world. Building a Soviet fleet was a national priority. When Germany invaded in 1941 and captured millions of soldiers, many sailors and naval guns were detached to reinforce the Red Army In February 194 6 the Red Fleet was renamed the Soviet Navy. The Soviet Navy was structured around submarines and small, maneuverable, tactical vessels. The Soviet shipbuilding program kept yards busy constructing submarines based upon World War II German Kriegsmarine designs, which were launched with great frequency during the immediate post-war years. Afterwards, through a combination of indigenous research and technology obtained through espionage from Nazi Germany and the Western nations, the Soviets gradually improved their submarine designs, though they initially lagged the NATO countries by a decade or two. Due to the USSR's geographic position, submarines were considered the capital ships of the Navy. It was submarines that could penetrate attempts at blockade, either in the constrained waters of the Baltic and Black Seas or in the remote reaches of the USSR's western Arctic. By the 1970s, Soviet submarine technology was in some respects more advanced than in the West, and several of their submarine types were considered superior to their American rivals.

Nuclear Propulsion

The Soviet Union soon followed the United States in developing nuclear-powered submarines in the 1950s. Stimulated by the U.S. development of the Nautilus, Soviet work on nuclear propulsion reactors began in the early 1950s at the Institute of Physics and Power Engineering, in Obninsk, under Anatoliy P. Alexandrov, later to become head of the Kurchatov Institute. In 1956, the first Soviet propulsion reactor designed by his team began operational testing. Meanwhile, a design team under Vladimir N. Peregudov worked on the vessel that would house the reactor. After overcoming many obstacles, including steam generation problems, radiation leaks, and other difficulties, the first nuclear submarine based on these combined efforts entered service in the Soviet Navy in 1958.

During the construction of the first November-class submarine, Russia also initiated a program in 1954 for building liquid-metal cooled submarine propulsion systems. Liquid metal cooled submarine reactors have been used by the Russian Navy. The technology was developed at the Institute of Physics and Power Engineering (IPPE) in Obninsk and used in two submarine classes: Project 645, a class in itself, and the Alfa class. Using liquid metal coolant was considered to have several advantages. It is more compact than pressurized water reactors, since it needs no moderator. No heavy pressure vessel is needed; it operates at higher temperatures and has therefore a higher thermal efficiency. The use of an intermediate reactor makes xenon poisoning less important. Refueling is faster, since the core is removed in a single operation. However, there are disadvantages: the melting point of the coolant is above room temperature, so the primary system must be kept heated at all times for the coolant to remain liquid. If not, the coolant will solidify and the cooling will be interrupted. The liquid metal coolant will gradually oxidize and the oxides must be removed regularly, to avoid blockage of the coolant flow through the core. The LMC reactor was first used in 1962 in a special version of a November-class submarine (Project 645, K-27), which used two RM-1 reactors with a capacity of 73 MWt each. The K-27 was re-fueled in 1967. However, it suffered a loss-of-coolant accident in 1968 in port when it was ordered to participate in a naval exercise at a time when the coolant needed to be cleaned of oxide impurities. During the exercise, these impurities blocked the entrance to the core of the port-side reactor and caused a LOCA, after which the submarine was laid up. In 1981, the free volume in the reactor and in the reactor compartment was filled with a conserving material and the submarine was sunk off Novaya Zemlya at 50 m.

At the height of the Cold War, approximately five to ten nuclear submarines were being commissioned from each of the four Soviet submarine yards. From the late 1950s through the end of 1997, the Soviet Union, and later Russia, built a total of 245 nuclear submarines, more than all other nations combined. They even built submarines of Titanium hull which came as a big surprise to the US and other nations.

Nuclear Powered Ice Breakers

At the same time, the Soviet Union started the work on the, nuclear propulsion' for the ice breakers. A nuclear-powered icebreaker is a nuclear-powered ship built for use in waters covered with ice. Nuclear-powered icebreakers are much more powerful than their diesel powered counterparts, and although nuclear propulsion is expensive to install and maintain, very heavy fuel demands and limitations on range can make diesel vessels less practical and economical overall, for these ice-breaking duties. During the winter, the ice along the Northern Sea Route varies in thickness from 1.2 to 2.0 meters (3.9 to 6.5 feet). The ice in central parts of the Arctic Ocean is on average 2.5 meters (8.2 ft) thick. Nuclear-powered icebreakers can force through this ice at speeds up to 10 knots (19 km/h, 12 mph). In ice-free waters the maximum speed of the nuclear-powered icebreakers is as much as 21 knots (39 km/h, 24 mph). The first nuclear powered ice breaker was named 'Lenin.' At its launch in 1957, the icebreaker NS Lenin was both the world's first nuclear-powered surface ship and the first nuclear-powered civilian vessel. Lenin was put into ordinary operation in 1959. Lenin had two nuclear accidents; the first in 1965, and the second in 1967. The second accident resulted in one of the three OK-150 reactors being damaged beyond repair. All three reactors were removed, and replaced by two OK-900 reactors; the ship returned to service in 1970. The Lenin was taken out of operation in November 1989 and laid up at Atomflot, the base for nuclear-powered icebreakers. Conversion to a museum ship was scheduled to be completed during 2005. Nuclear-powered icebreakers have been constructed by the only country USSR and later Russia primarily to aid shipping along the Northern Sea Route in the frozen Arctic waterways north of Siberia. In all, ten civilian nuclear-powered vessels have been built in the USSR and Russia. Nine of these are icebreakers, and one is a container ship with an icebreaking bow. The icebreakers have also been used for a number of scientific expeditions in the Arctic.

Since 1989 the nuclear-powered icebreakers have also been used for tourist purposes carrying passengers to the North Pole. Each participant pays up to US$ 25,000 for a cruise lasting three weeks. During the summer of 1993, the NSYamal was used for three tourist expeditions in the Arctic.

The NS Yamal has a separate accommodation section for tourists. The nuclear-powered icebreaker 50 Let Pobedy (known in English as the 50 Years of Victory) contains an accommodation deck customized for tourists. Some ships carry one or two helicopters and several Zodiac boats. Radio and satellite systems can include navigation, telephone, fax, and email capabilities. Most nuclear-powered icebreakers in the Russian service today have a swimming pool, a sauna, a cinema, and a gymnasium. In the restaurants aboard there is a bar and facilities for live music performances. Some also have a library and at least one has a volleyball court. Russia is planning to start building new icebreakers. "It is important to not only use the existing fleet of icebreakers, but also to build new ships, and the first nuclear icebreaker of a new generation will be built by 2015. This should be an icebreaker capable of moving in rivers and seas"

The Typhoon, with a submerged displacement of more than 48,000t, is the world's biggest submarine class. It is a nuclear-powered submarine equipped with ballistic missiles. Dmitry Donskoy, the first of the six submarines in the class, was commissioned in 1981 and is still in active service with the Russian Navy. The largest submarines ever constructed, they measure 175 meters long with a submerged displacement of 48,000 tons. The Typhoon class is capable of staying submerged for up to three months at a time.

NUCLEAR PROPULSION IN UK

At the beginning of the 20th century, the idea of submarine warfare was considered by senior personnel in the Admiralty to be "underhand, unfair and damned un-English" However, those in favor of experimenting with submarine technology eventually won the argument, and the Royal Navy launched its first submarine, Holland 1, in 1901.

The Submarine Service proved its worth in World War I, where it was awarded five of the Royal Navy's 14 Victoria Crosses of the war, the first to Lieutenant Norman Holbrook, Commanding Officer of HMS B11.

The Royal Navy Submarine Service had 70 operational submarines in 1939. Three classes were selected for mass production, the seagoing "S class" and the oceangoing "T class" as well as the coastal "U class". All of these classes were built in large numbers during the war.

The main operating theatres for British submarines were off the coast of Norway, in the Mediterranean, where a flotilla of submarines successfully disrupted the replenishment route to North Africa from their base in Malta, as well as in the North Sea. As Germany was a Continental power, there was little opportunity for the British to sink German shipping in this theatre of the Atlantic.

From 1940, submarines were stationed at Malta, to interdict enemy supplies bound for North Africa. Over a period of three years, this force sank over 1 million tons of shipping, and fatally undermined the attempts of the German High Command to adequately support Rommel. Rommel's Chief of Staff, Fritz Bayerlein conceded that "We would have taken Alexandria and reached the Suez Canal, if it had not been for the work of your submarines". 45 vessels were lost during this campaign, and five Victoria Crosses were awarded to submariners serving in this theatre.

In addition, British submarines attacked Japanese shipping in the Far East, during the Pacific campaign. The Eastern Fleet was responsible for submarine operations in the Bay of Bengal, Strait of Malacca as far as Singapore, and the western coast of Sumatra. Few large Japanese cargo ships operated in this area, and the British submarines' main targets were small craft operating in inshore waters. The submarines were deployed to conduct reconnaissance, intercept Japanese supplies travelling to Burma, and attack U-boats operating from Penang. The Eastern Fleet's submarine force continued to expand during 1944 and by October 1944 had sunk a cruiser, three submarines, six small naval vessels, 40,000 tonns of merchant ships, and nearly 100 small vessels. In this theatre, the only documented instance of a submarine sinking another submarine while both were submerged occurred; HMS Venturer engaged the U864 and the Venturer crew manually computed a successful firing solution against a three-dimensionally maneuvering target using techniques which became the basis of modern torpedo computer targeting systems.

By March 1945, British boats had gained control of the Strait of Malacca, preventing any supplies from reaching the Japanese forces in Burma by sea. By this time, there were few large Japanese ships in the region, and the submarines mainly operated against small ships which they attacked with their deck guns. The submarine HMS Trenchant torpedoed and sank the heavy cruiser Ashigara in the Bangka Strait, taking down some 1,200 Japanese army troops. Three British submarines (HMS Stonehenge, Stratagem, and Porpoise) were sunk by the Japanese during the war.

The Royal Navy had been researching designs for nuclear propulsion plants since 1946, but this work was suspended indefinitely in October 1952. The UK's Naval Nuclear Propulsion Programme, NNPP is closely tied to the United States. The United States supplied the UK with its first naval nuclear reactor and has supplied fissile material for core fabrication. US assistance to the UK NNPP has been part of a wider programme of military nuclear cooperation that dates back to the 1940s Manhattan project. Cooperation abruptly ended in 1946 when the US Congress passed the Atomic Energy Act (the McMahon Act) that severely limited the transfer of restricted nuclear information and materials to any other state, causing a major rift with its wartime ally in London. Negotiations to reestablish exchange of military atomic defense information, materials and technology began after an independent UK nuclear weapons programme tested its first atom bomb in 1952 and hydrogen bomb in 1957. Surprisingly, every country made a nuclear weapon, before making nuclear reactors for propulsion. Britain's nuclear dependence on the United States was then cemented in the 1958 Mutual Defense Agreement (MDA) and five years later the 1963 Polaris Sales Agreement. The MDA allowed for the exchange of naval nuclear propulsion technology between the United States and UK. The UK had already initiated an indigenous NNPP in 1956 a year before the decisive H-bomb test, but by then the United States programme was already developing its fifth generation naval reactor. After a series of exchange visits in 1957-58, the United States agreed to supply the UK with one complete submarine nuclear reactor plant. This was the latest S5W pressurised water reactor (PWR) design for the US Skipjack-class submarine. The S5W went on to power a total of 98 US submarines, including different classes of SSNs and the Polaris SSBN fleet. The agreement reached under the auspices of the MDA involved the transfer of reactor technology and manufacturing expertise from Westinghouse in the United States to Rolls Royce in the UK. This enabled the UK to deploy its first nuclear-powered submarine, HMS Dreadnought, several years earlier than originally envisaged. HMS Dreadnought was launched in 1960 and commissioned into service in 1963. Under the terms of the agreement, cooperation on nuclear propulsion came to an end in mid-1963; exactly one year after the UK's S5W plant became operational. NNPP cooperation was terminated as a condition of the transfer, in order to ensure future UK operational, design and safety independence. On 3 March 1971, Dreadnought became the first British submarine to surface at the North Pole. In 1973 she took part in the Royal Navy's first annual Group Deployment, when a group of warships and auxiliaries would undertake a long deployment to maintain fighting efficiency and "show the flag" around the world. In the United Kingdom, all former and current nuclear submarines for their British Royal Navy (with the exception of three: HMS Conqueror, HMS Renown and HMS Revenge) have been constructed in Barrow-in-Furness, where construction of nuclear submarines continues. Conqueror is the only nuclear-powered submarine ever to have engaged an enemy ship with torpedoes, sinking the cruiser ARA General Belgrano with two Mark 8 torpedoes during the 1982 Falklands War.

The Royal Navy Submarine Service is the submarine element of the Royal Navy. It is sometimes known as the Silent Service, as the submarines are generally required to operate undetected. The service operates seven fleet submarines (SSNs), of the Trafalgar and Astute classes (with four currently planned or under construction), and four ballistic missile submarines (SSBN), of the Vanguard class. All of these submarines are nuclear powered. The service also owns the LR5 Submarine Rescue System.

The UK's first indigenous plant and core (PWR1) based on the S5W Westinghouse design was built by Rolls Royce and deployed in the attack submarine HMS Valiant in 1966. Three sets of cores were developed for PWR. Core 1 powered the UK's Valiant-class SSNs and Resolution-class SSBNs that carried the US Polaris strategic weapon system, core 2 powered the Churchill class SSNs, and core 3 powered the Swiftsure and Trafalgar-class SSNs. The current PWR2 reactor was designed for the UK's Vanguard-class SSBNs that entered service in the 1990s to replace the Resolution-class and carry the US Trident strategic weapon system. Design work began in 1977 and the first PWR2 reactor was completed in 1985 with testing beginning in August 1987. The current PWR2 core design, 'Core H', has been designed to last the lifetime of the reactor (25-30 years) thereby eliminating costly mid-life reactor refueling. Core H fuels the new Astute-class SSNs and the four Vanguard-class SSBNs were fitted with the core during their long overhaul and refueling refits between 2002 and 2012.

UK reliance on US NNPP expertise is set to deepen with the development of a new PWR3 reactor for the planned Successor SSBN announced in the coalition government's Submarine Initial Gate Parliamentary Report in May 2011. MoD's Defense Board said the PWR3 would be 'based on a modern US plant and US support', will also be provided independent peer review of the UK's NNPP capability and will help to optimize its PWR3 concept design. In 2012 the US Navy said: 'Naval Reactors is providing the UK Ministry of Defense with US naval nuclear propulsion technology to facilitate development of the naval nuclear propulsion plant for the UK's next generation SUCCESSOR ballistic missile submarine'. It has been suggested that MoD has been 'given visibility of the S9G reactor design that equips the US Navy's latest Virginia-Class nuclear-powered attack submarines'. PWR3 will in all likelihood power the new Successor flotilla. The US-UK MDA is renewed every 10 years. In the latest update negotiated and agreed in July 2014, Article III of the treaty was modified to authorize transfer of new reactor technology, spare parts, replacement cores and fuel elements. The original text of the treaty referred to the transfer by sale of only one complete submarine nuclear propulsion plant (the original S5W). The UK has also been dependent upon the United States for special nuclear materials (highly enriched uranium and plutonium) and tritium gas for its nuclear warhead and nuclear propulsion programmes. The UK initially obtained HEU for its military programme from its gas diffusion plant at Capenhurst. Capenhurst was established as the sole UK uranium enrichment site for both civil and military applications. This began in 1952, but production of HEU (high enriched uranium) for military purposes ended a decade later in 1962. Since then the UK has received HEU for both its warhead programme and NNPP through exchanges of special nuclear material with the US Department of Energy under the MDA. The UK imported natural uranium ore concentrate from the United States, Australia, South Africa, Namibia, Belgian Congo and Canada and converted it into uranium hexafluoride (UF6) at the UK's Springfields site in Lancashire. It was then shipped to the United States for enrichment at the US Portsmouth Gaseous Diffusion Plant in Piketon, Ohio. HEU enriched, between 93-97% in 235U was transported back to the UK by military aircraft.

IPFM –International Panel on Fissile Materials, a nongovernmental organization that tracks HEU and plutonium worldwide estimates, that the UK is estimated to have received more than half of its HEU supply from the United States with an estimated transfer of at least 14 tonnes. The UK appears to have shown little interest in the development of LEU (low enriched uranium) cores for submarine nuclear reactors. It remains tied to the US NNPP programme for materials, design and support under the auspices of its nuclear weapons programme; it also remains heavily dependent upon continued US patronage. Unless and until the US NNPP opts for an LEU propulsion plant and is prepared to share (or perhaps co-develop) such a plant with the UK, or until the UK terminates its nuclear-powered submarine programme, the UK looks set to power its submarine flotilla with current and future iterations of its HEU-fuelled PWR3. UK nuclear weapons policy is, like many other policies, subject to political, technical and organizational resistances to changes in policy and practice. The challenge of prevailing conceptions of what constitutes an 'effective' nuclear deterrent threat; in exchange for ambiguous non-proliferation benefits is not an incentive. The conservatism of nuclear policy communities has been well documented. This applies equally to the prospect of conversion from HEU to LEU; non-proliferation concerns alone are unlikely to be an incentive for a transition in the UK NNPP programme to LEU.

The UK's nuclear-powered submarines are currently defueled and refueled at the Devonport Royal Dockyard at least once during their service life. Rolls Royce's current 'Core H' is designed to last the service life of the new Astute-class SSN thereby eliminating the need for mid-life refueling. All four Vanguard-class SSBNs were fitted with a 'Core H' as part of their planned Long Overhaul Period (Refuel) between 2002 and 2012. However, in March 2014 MoD revealed a breach in the fuel cladding of the PWR2 prototype test reactor at Dounreay that allowed low-level radiation to leak into its sealed cooling circuit. As a result a decision was taken to replace the core in HMS Vanguard again during its next planned maintenance visit to Devonport in 2015.

Defueling of submarines is currently carried out from 'a mobile Reactor Access House (RAH) which traverses the dry dock and is positioned above the reactor compartment (RC) of the submarine. Access to the RC is gained by cutting holes in the submarine pressure hull directly above the reactor pressure vessel (RPV). The spent Fuel Modules (FM) and Neutron Sources (NS) are then raised from the RPV and temporarily parked in shielded storage boxes within the shielded tank. The FM/ NS is then lifted into a shielded transport container before removal from the RAH to another facility within the Dockyard for onward processing. Irradiated fuel removed from submarines is moved to a storage facility for the temporary storage prior to consignment to Sellafield.

Future Submarines in the UK

A total force of seven Astute class fleet submarines is planned. As of April 2016, the first three boats are in commission and in service, while boats four to six are in various stages of construction. HMS Ambush, the Royal Navy's newest nuclear attack submarine, is one of the most sophisticated and powerful vessels of her type ever built.

News---"The giant Astute-class sub, which was launched today (Nov 2012), is so hi-tech she doesn't even need a periscope. Her crew instead using a digital camera system to see above the surface when she is submerged she has enough nuclear fuel to carry on cruising for up to 25 years non-stop - giving her huge tactical flexibility. Her nuclear reactor is so powerful her range is only really limited by the need for maintenance and resupply. Astute-class submarines are the largest, most advanced and most powerful in the history of the Navy, boasting world-class design, weaponry and versatility. HMS Ambush can travel over 500 miles in a day, allowing them to be deployed anywhere in the world within two weeks. The vessel is also one of the quietest sea-going vessels built, capable of sneaking along an enemy coastline to drop off Special Forces or tracking a boat for weeks. Foreign forces will find it almost impossible to sneak up undetected by her incredibly powerful sonar equipment that can hear halfway around the world".

Boat number seven was confirmed in the October 2010. Strategic Defense and Security Review and long-lead items have been ordered. The Astute-class submarine is the largest nuclear fleet submarine ever to serve with the Royal Navy, being nearly 30% larger than its predecessors. Its power plant is the Rolls Royce PWR2 reactor, developed for the Vanguard-class SSBN. The submarine's armament consists of up to 38 Spearfish torpedoes and Tomahawk Block IV land-attack cruise missiles.

A successor to the Vanguard class of submarines is in its early stages, with 'Main Gate' decisions to be made in 2016. The programme will seek to replace the current Vanguard-class ballistic missile submarines starting sometime during the mid-to-late 2020s.

NUCLEAR PROPULSION IN FRANCE

The day of 29th March 1967 was a milestone in the history of the French Navy and the military port of Cherbourg in Normandy. After a three-year build period, France's first Ship Subsurface Ballistic Nuclear (SSBN) submarine Le Redoutable was launched. The tall figure of the French President Charles de Gaulle dominated the many political and military personalities who attended the event. Francis Perrin, High Commissioner of the French Atomic Energy Commission (CEA, Commissariat à l'Energie Atomique) was among them. He succeeded Frédéric Joliot-Curie in this position and he was also the son of another Nobel Prize winner, Jean Perrin. In his speech, the Defense Minister Pierre Messmer paid tribute to all those, from scientists to engineers and workers, who made the major achievement possible and concluded by expressing his confidence that Le Redoutable will be an instrument of security, independence and peace. As de Gaulle to pressed on a green button, the 128 meters of Le Redoutable slowly slid down the ramp of Hold 3, the national anthem La Marseillaise was played and the submarine weighing more than 9,000 tons gently made her first contact with seawater. This nuclear-powered ballistic missile-carrying submarine (SSBN) used a high strength elastic steel hull, enabling her to dive to 300 meters.

France, one of the five Permanent Members of the United Nations Security Council, joined the USA, the USSR and the United Kingdom in the small club of countries equipped with nuclear submarines. This was the beginning of a new chapter in the already long histories of French submarines and French nuclear technology.

Brief history of French submarines

Denis Papin (1647-1712), inventor of the steam digester, forerunner of the pressure cooker and of the steam engine was among the few French who designed the first primitive submarines in 1800 in the river Seine near Rouen then in 1801 near Le Havre and Brest, a submarine called Nautilus is tested. The American Robert Fulton had suggested to the French Directory that his machine could help France break the naval blockade of the French coasts enacted by the British government. Successful as they were, the first real tests did not convince Napoleon Bonaparte who had overthrown the Directory in the meanwhile. During attacks in the Bay of Isigny in Normandy, the British ships are warned by spies and fled. The Nautilus is too slow to chase them.

In August 1832, the inhabitants of the island of Noirmoutier, on the Atlantic Ocean, look in awe at a fishing boat with a crew of three men diving five meter deep and resurfacing after about twenty minutes. The submarine is 3.2 meters long and displaces about six tons when submerged.

She has three sets of duck-foot paddles and a large rudder. The designer is Brutus de Villeroi (1794–1874) who would repeatedly fail to get the interest of the French Navy. He would eventually immigrate to America where he would design the USS Alligator that would be used during the Civil War. In fact de Villeroi was not an engineer, he taught mathematics and drawing at a school in Nantes, the city of a young man called Jules Verne, the future author of Twenty Thousand Leagues under the Sea.

Prosper Payern (1806-1886) was a medical doctor with surprising ideas such as digging a railway tunnel under the English Channel. He also succeeded in using lime to absorb carbon dioxide and using potassium or manganese superoxide to regenerate oxygen in diving bells. In July 1846, Payern presented to 20,000 Parisians the first workable and efficient submarine ever, propelled mechanically by a screw propeller. Shaped like an egg, the Belledonne was 9 meters long, 2.8 m wide and weighed 10 tons. She would be used for marine works during six years, participating among others to the digging of the channel in the Cherbourg port.

Between 1863 and 1867, all efforts to develop Le Plongeur, a submarine using a compressed-air engine could not solve the problems of stability. The 23 tanks holding air took up a huge amount of space and required the submarine to be very large. The next major step would be the Gymnote (hull number Q1) designed by Henri Dupuy de Lôme (1816-1885) and Gustave Zédé (1825-1891) who advised Jules Verne to design - on paper only - Captain Nemo's Nautilus. Built in Toulon and launched in 1888, the Gymnote was propelled by a multipolar electric engine of 60 HP with a battery made of 564 accumulators weighing 9,840 kg. With a crew of five, the Gymnote could travel at a constant depth below the surface and was fitted with a periscope, an electric gyroscope and two torpedoes. Though the speed was only eight knots on the surface and four knots when submerged, in 1890 Le Gymnote was the first submarine to force a blockade by diving under the keel of a battleship without being seen. The electric and mechanical systems gave such satisfaction that they would be used for a larger boat of the same type named after Gustave Zédé. However, neither Le Gymnote nor Le Gustave Zédé were fit to engage in combats and both were unable to go far from the base.

The twentieth century saw the opening of the era of the submersibles, able to travel long distances on the surface and diving only for combat. Maxime Lauboeuf (1864-1939) was the father of modern French submarines and many of his innovations were adopted in other countries. In 1904, Laubœuf built le Narval. Equipped with external ballasts, it was the first submarine with mixed propulsion: a steam engine when on the surface and an electric engine when submerged. Most submarines that were built in the twentieth century would use this concept: when resurfacing, the batteries are recharged by a generator powered by the diesel engine. The

First World War dramatically demonstrated the crucial role that submarines could play in warfare. Submariners started dreaming of a submersible that would use a single engine and could dive for extended periods of time without resurfacing. Nuclear propulsion would be the answer.

Development of the French nuclear technology

Between the two World Wars several French scientists such as Frédéric and Irène Joliot-Curie made a major contribution to nuclear physics. In February 1939, Frédéric Joliot-Curie, Hans von Halban and Lew Kowarski carried out an experiment at the College de France showing that the fission of a nucleus of uranium yields neutrons and demonstrating the possibility to create chain reactions. In May 1939 three patents were submitted via the public research organization CNRS (Caisse Nationale de Recherche Scientifique): two were for the production of nuclear energy and the third for the enhancement of explosive charges. These patents and two more associating Francis Perrin submitted in 1940 would later pose challenging intellectual property problems in the USA. In July 1939, experiments to produce energy from the atom started on a somewhat larger scale at the Laboratory of Atom Synthesis in Ivry-sur-Seine near Paris, in parallel with the work of Francis Perrin and Fermi's team in the US. The objectives were to understand the propagation of neutrons and to build a prototype producing energy from chain reactions, or at least to demonstrate its feasibility. On 3rd September 1939, France and Britain declared war on Germany. The research project was placed under the Ministry of Armaments and classified after Joliot-Curie had explained to the newly appointed Minister Raoul Dautry the potential of nuclear energy for civil and military applications.

In 1940, the invasion of France appeared imminent. Joliot-Curie decided to hide the stock of uranium oxide: seven tons were shipped to Morocco and nine tons were loaded in a wagon that was found only at the end of the war in Le Havre. As for heavy water (deuterium oxide) of which large quantities are needed to slow down the chain reaction, France only had about 50 grams. The only existing stock of heavy water was held by NorskHydro in Rjukan, Norway. Before the Nazis invaded this neutral country, Raoul Dautry sent secret agents to negotiate with Norway who consented to give the stock. Though the Germans controlled all roads and communications, the 185 kg of precious liquid would eventually reach Paris via Edinburg on 16 March 1940. In May, the French army having been severely defeated by the Germans. Joliot-Curie instructed von Halban et Kowarski, both of Jewish descent, to leave for England with the stock of heavy water, one gram of radium and sealed letters containing the recent research results and the next plans. On 18 June 1940 the two scientists left Bordeaux on the English coal ship Broompark. Though nuclear research was more probably more advanced in France than anywhere else, it was totally suspended until the end of the Second World War. With the exception of Joliot-Curie who stayed in occupied France, the main figures would cooperate with the allies in the UK (MAUD Commission) then in Canada in a laboratory participating in the Manhattan project.

Two months after Hiroshima and Nagasaki, General de Gaulle created the Atomic Energy Commission (Commissariat à l'Energie Atomique, CEA) in October 1945. Joliot Curie was named High Commissioner and undertook the design of Zoé, the first French nuclear reactor, with no enriched uranium available. The reactor was a pool-type design, with five tons of heavy water moderator surrounded by a two meter-thick graphite-concrete wall. The core, immersed in the pool, consisted of 60 aluminum clad vertical rods containing three tons of uranium oxide pellets, controlled by cadmium rods. Built in less than 16 months, Zoé went critical on 15 December 1948. Its power was very low (Kowarski facetiously suggested to call it FLOP for French Low Output Pile) but it was to become a workhorse that would be used till 1976.

Not everyone rejoiced at the success of Zoé. On 27 December 1948 the editorialist of the New York Herald Tribune wrote: The existence of a French reactor is a true threat to the measures that English-speaking nations thought adequate. For many, the threat is even more serious given the fact that the head of the French work, Dr Frédéric Joliot-Curie is an avowed communist.

A second heavy water reactor, EL2, reached criticality in 1952 in the new nuclear research center at Saclay. That same year, a five-year plan for nuclear activities was voted in the hope of remedying the energy deficit of post-war France. The NUGG (Natural Uranium Graphite Gaz) technology was adopted: the reactor was fueled with natural uranium metal, graphite moderated and cooled by carbon dioxide. On the site of Marcoule in the Rhone valley, reactors G1 (46 MW), G2 (250 MW) and G3 (250 MW) reached criticality in 1956, 1958 and 1959 respectively. These were the first steps on the path that would lead to the vast and successful nuclear power generation programme of the 1970s.

In the late 1940s, the military use of atomic energy had been the object of a heated debate. One of the leading opponents was none other than Frédéric Joliot-Curie who had initiated the Stockholm Appeal calling for an absolute ban on nuclear weapons. Joliot-Curie was dismissed from the CEA in April 1950. The successive governments of the weak Fourth Republic were unable to reach a decision until the end of 1956 when a mixed CEA-military Committee for the Military Applications of Atomic Energy was formed. Pierre Mendès-France had stated: «without the bomb, one does not have a say". His government approved an outline for a strategic nuclear bomb and the design of vehicles of delivery. When de Gaulle returned to power in May 1958, he accelerated the French nuclear program insisting that the country should not depend on foreign technologies.

Because he was a Jew, Bertrand Goldschmidt also had to flee France in 1940. Though the U.S. government did not permit the participation of French scientists, an exception was made for Goldschmidt who participated in the development of the PUREX process to separate plutonium and uranium. Back in France in 1946, he obtained the first four milligrams of plutonium from the spent fuel of the Zoé reactor in November 1949. After the construction in 1954 of a demonstration unit using the PUREX process in Fontenay, a full scale unit (UP1) was built in 1958 in Marcoule to produce plutonium from reactors G1, G2 and G3.

13 February 1960: Gerboise Bleue (Blue desert rat) explodes in the Sahara desert with an explosive yield of 70 kilotons. After the USA, the USSR and the United Kingdom, France has become the fourth country having the atomic bomb.

The nuclear propulsion research programmes till 2000

The Q 244 project that started in 1955 was the first French attempt to build a nuclear propulsion reactor. The technology would be what was known best and could realistically be achieved with what was available: natural uranium and heavy water. This would require a submarine larger than the largest then, the Surcouf with its 3,500 tons. The length was 113.7 m, the width 11.7 m and it would displace 6,500 tons. Among many technical problems, the major one certainly was that the nuclear reactor would simply not fit in this hull. The project was abandoned in 1958 and France turned to the USA for help. Out of their wish list, the French delegation only obtained 20 tons of enriched uranium, in spite of the strong opposition of Admiral Rickover, the crusty father of the US nuclear Navy. When asked during a Congress hearing if there was any chance that France would succeed in developing a propulsion reactor, the Admiral replied 'not the slightest'. From which it was concluded that the supply agreement could be signed. One of the conditions was that the enriched uranium would only be used for a research reactor on the land.

The first nuclear propulsion reactor was built in the new nuclear research center of Cadarache, 70 km North of Marseille, destined to be the new test bench of the CEA for various advanced nuclear technologies. Since the agreement with the Americans excluded any technology transfer, the small Nuclear Propulsion Group GPN (Groupe de Propulsion Nucléaire) started from scratch and miniaturized the components of the reactor. On 11 September 1959, the GPN received the requirements from the Navy: a concise one-page document accompanied by a request to receive the preliminary project document two months later. The Navy received it on 17 November 1959. To build a submarine on land, the first step was to build a pool, place a cylinder in it to eventually fit the reactor in the cylinder. Around these the control and command systems were installed as well as very realistic facilities to dissipate power. The Prototype on Land PAT (Prototype A Terre) used pressurized water and the enriched uranium from the USA, which would allow a more compact reactor. The circuit between the reactor vessel and the steam generators would form two loops.

On 9 April 1962 the low power reactor Zirconium Uranium Alloy Azur (Alliage Zirconium Uranium), also in Cadarache, reached criticality. It would help defining the neutronics of the core and testing the primary protection. The reactor was embarked in February 1963 and went critical on 14 August 1964. GPN and the vendors met the deadline and the budget.

From mid-October to mid-December 1964, the reactor made a smooth virtual tour of the world. For about ten years, the PAT would be instrumental in refining the design and operation of loop reactors that would equip the first generation of SSBN. Before it was stopped in 1992, the PAT went critical more than 3,500 times and was used to train nearly 3,000 submariners.

The first uranium enrichment plant started in 1965 next to the G1, G2 and G3 NUGG reactors in Pierrelatte. France was now independent for its supply of enriched uranium. The PAT was the model for the first six SSBNs.

Hull No	Name	Laid down	Launch	Completed	Decommission
S611	*Redoutable*	1964	29 Mar 1967	1 Dec 1971	1991
S612	*Terrible*	1967	12 Dec 1969	1 Jan 1973	1996
S610	*Foudroyant*	1969	4 Dec 1971	6 Jul 1974	1998
S613	*Indomptable*	1971	17 Aug 1974	23 Dec 1976	2003
S614	*Tonnant*	1974	17 Sept 1977	3 Apr 1980	1999
S615	*Inflexible*	1980	23 Jun 1982	1 Apr 1985	2008

With its PWR reactor, two steam turbines and a turbo-reducing group the whole propulsion system of Le Redoutable delivered 16.000 HP. During her lifetime Le Redoutable would spend 90,000 hours underwater and would travel 1,270,000 Kilometers. L'Inflexible, the latest SSBN in this class, would incorporate significant improvements regarding nuclear safety, propulsion and acoustic stealth.

The Navy being satisfied with the first SSBNs in operation, the time had come to build attack submarines (SSAs). These would use the same nuclear propulsion technology but for the more affordable hull of the Agosta class of diesel-electric submarines. The reactor would therefore need to be much smaller and lighter. It also had be more silent, faster to install on board with less interference between the installation of the nuclear steam supply system and the general wiring of the boat. To meet these objectives the construction of a new prototype started in Cadarache in 1970: the Advanced Boiler Prototype CAP (Chaufferie Avancée Prototype). A major innovation was introduced: the steam generator was placed directly above the core of the reactor, so that there were no more circuits between the two. The benefits were dramatic:

• The primary block was very compact resulting in a massive weight reduction from 700 to 400 tons. To date these SSAs are still the most compact in the world (2,300-2,700 tons).

• Up to a significant power, the coolant was naturally circulating and thus totally silent. At higher speed the reactor was still very silent, as only low-power pumps were required and the vessel-steam generator block filtered the noise.

• The installation time (boiler and submarine) was drastically reduced.

The new boiler K48 (48 MW) derived from the CAP was installed in the Rubis class of SSAs.

Hull number	Name	Laid down	Launched	Completed
S 601	*Rubis*	1976	7 Jul 1979	23 Feb 1983
S 602	*Saphir*	1979	1 Sept 1981	6 Jul 1984
S 603	*Casabianca*	1981	22 Dec 1984	13 May 1987
S 604	*Émeraude*	1983	12 Apr 1986	15 Sept 1988
S 605	*Améthyste*	1984	14 May 1988	20 Mar 1992
S 606	*Perle*	1987	22 Sept 1990	7 Jul 1993

These boats can dive to 300 meters, thanks to the high elastic modulus of the 80 HLES steel of the hull. The massive sonar domes are made of composite material. They have two crews of 70 men and can navigate 220 days in a year.

Unfortunately, the noise levels turned out to be higher than expected and a silencing program needed to be implemented. This came with Améthyste, the name of another gemstone but also a French acronym for Improvement Tactic Hydrodynamics Silence Tansmission Tapping (Amelioration Tactique HYdrodynamique Silence Transmission Ecoute). The sonar and the electronics were upgraded and the shapes of the bow and hull were changed. These changes having proved successful on the fifth and sixth hulls (Améthyste and Perle), the first four boats were modified to the same standards between 1989 and 1995.

The cat and mouse game between submarines and surface combatants that began with the German U-boats during the First World War and continued through the Cold War era is endless. The evolution of detection methods implied further reducing acoustic signatures to much lower levels. In the early 1980s the French Navy planned to replace the first generation of SSBN by boats that would be much more silent but also to replace the ageing aircraft carriers Foch and Clemenceau with a nuclear aircraft carrier. In order to save on development cost Technicatome suggested the K15 PWR concept that would be almost the same for both applications and would draw on the previous successes. Starting in 1984, a nearly total overhaul of the CAP was undertaken in Cadarache. Only the original vessel was kept for the New Generation Reactor RNG (Réacteur Nouvelle Génération), the first reactor using digital control systems to drive the control rods and ensure safety. With the RNG major advances were made in the development of ever more efficient fuels but also of hydraulic systems to remove some mechanical components. Significant progress was also made in the field of corrosion.

The construction of Le Triomphant started in June 1989 in the Cherbourg naval dockyard of the Naval Construction Directorate DCN (Direction des Constructions Navales). She had an anechoic coating designed to absorb sonar waves and it was the first submarine equipped with a ducted propeller (propeller pump) to avoid cavitation. To avoid propagation of the inevitable noise outside the hull, the rigid connections between the noisy hardware and the hull were replaced by cradles connected to the hull via suspensions filtering the vibrations. On these cradles, each engine, each pipe, each electric cable was in its turn suspended or placed on other suspensions filtering noise and vibrations. The noisy machines were covered with insulating caps.

Le Triomphant was launched in 1994 and entered active service in 1997. She displaces 14 335 t, her speed is approximately 25 knots and she can dive to a depth of 400 meters. The special steel of her hull withstands 100 kg/mm. Most of all, she produces approximately one thousandth of the detectable noise as compared to Le Redoutable-class boats and she is ten times more sensitive in detecting other submarines. The K15 boiler of 150 MW built by Areva Technicatome has a capacity of 41,500 HP (30,519 kW) and an autonomy of 6 years without refueling. These are the most expensive submarines ever produced.

Today

The ten French military submarines are all nuclear submarines. They are the main component of the Oceanic Strategic Force FOST (Force Océanique Stratégique) of France's nuclear deterrent strike force (Force de

Frappe).

Ile Longue near Brest at the Western tip of France is the home port of the four New Generation SSBNs Le Triomphant, Le Téméraire, Le Vigilant and Le Terrible. At any time, at least one of them is patrolling with sixteen strategic missiles M51, conventional torpedoes and Exocet anti-ship missiles. These are expected to be in service at least until the mid-2030s.

The military harbor of Toulon on the Mediterranean Sea is home to the six SSAs of the Rubis class: Rubis, Saphir, Casabianca, Émeraude, Améthyste and Perle. They are armed with F17 torpedoes and SM-39 Exocet anti-ship missiles and equipped with a multifunction sonar, a very low frequency towed sonar, a radar detector ARUR 13 and a satellite communications system. The initial plan was to keep these attack submarines in service for 25 years but it was later decided to keep them one decade longer. From 2018 they will be gradually replaced by the SSAs of the Suffren class.

Also based in Toulon, the aircraft carrier Charles de Gaulle was commissioned in 2001. She is equipped with two K15 reactors and four diesel-electric engines developing 40,230 HP each while driving two shafts. Her speed is 27 knots. She fields the powerful naval variant of the Dassault Rafale (known a "Rafale M") and can also sport a variety of other specialized aircraft and helicopters as needed.

With a total of 12 nuclear propulsion reactors in service, it is necessary to have onshore facilities that simulate onboard facilities. This is why the Test Reactor RES (Réacteur d'Essais) designed by the CEA and to be operated by Areva TA in Cadarache was built from 2003. RES will be used for the qualification of calculation codes, new fuels and equipment modifications on existing and future onboard nuclear boilers. The overall objective is to maintain and continuously improve nuclear safety, cost and performance. RES will also be used to train the crews; it also has a pool in use since 2005 to store and analyze spent fuel elements. This is a reactor that is essentially a K15 reactor with a lot of instrumentation in its core for neutron mapping or monitoring of the neutron flux in the core in real time. The reactor was expected to go critical in 2014 but it is likely that it will not happen before 2017.

Tomorrow : Barracuda

Nuclear propulsion is now such a mature technology that it is no longer necessary to validate fully new concepts as was the case with the PAT, the CAP and finally the RNG prototype that was stopped in 2005. The challenge now lies in the ability to focus and validate parts of the concept, especially the core and nuclear fuel formulations. The future submarines, such as those of the Barracuda program do not need a prototype reactor but a test bench such as RES.

From 2017 to 2030, the current fleet of SSAs will be replaced by a new generation with a reinforced strategic vocation. Both battle ships and instruments of power, they will carry naval cruise missiles MdCN (Missile de Croisière Naval) with an in depth strike capability on the earth and increased resources to support special operations. In order to make this possible, space for ten additional persons has been created inside the hull. External equipment module, to be used by the commandos, may be mounted outside. This location may also be used to transport UUV systems.

The Barracuda program was launched in 2006. In order to better manage the interface between the boat and the boiler room, the CEA and the French Defense Procurement Agency DGA (Délégation Générale à l'Armement) formed an integrated project management team. An added advantage is that the prime contractors, Areva TA for the nuclear part and the Directorate for Naval Constructions Services DCNS (Direction des Constructions Navales Services) have to deal with a single customer.

The nuclear propulsion system will be a new hybrid design providing electric propulsion for economical cruise speeds and turbo-mechanical propulsion for higher speeds. The reactor will be an improved K15, 50 MW used for "Le Triomphant" class and the de Gaulle. The entire French nuclear fleet will therefore use the same type of reactor, simplifying procurement and maintenance. Thanks to new developments the Barracuda will reach at least 25 knots submerged. Other improvements relate to safety with a design taking into account the possibility of severe accidents. Exposure to radiation protection will be further reduced. For example, thanks to automated systems, the reactor operator station will now be located at the front of the boat. The new fuel that is used requires refueling every ten years instead of the current seven years. Another novelty is that the steam generated by the reactor is used not only for the turbine directly driving the propeller directly but also for turbo generators that can power electric motors. The results are more energy efficient and very quiet boats.

Larger, more powerful and quieter than the Rubis class, these new generation SSAs will be 99.5 meters long and will displace 5,300 tons. Highly automated, they will be armed with a crew of 60 men (two teams will take turns as is currently the case), which may be partially feminized if the experiment conducted in 2017 on SSBNs is conclusive. The food storage capacity will enable the crew to live in complete self-sufficiency for 70-90 days against 45-60 days now. The first new Barracuda SSN, Suffren, is on schedule to be launched in early 2017. The next boatss in the Suffren-series will be Duguay-Trouin, Tourville, De Grasse, Rubis and Casabianca.

NUCLEAR PROPULSION IN CHINA

Submarine warfare is regarded as a vital part of PLAN's coastal defense doctrine. Large numbers of conventionally powered submarines have therefore been in service, and this force makes up the bulk of the PLAN's submarines, making it the second largest submarine force in the world today.

The People's Liberation Army Navy (PLAN), also known as the PLA Navy, is the naval warfare branch of the People's Liberation Army, the armed wing of the Communist Party of China and by default, the national armed force of the People's Republic of China. The PLAN can trace its lineage to naval units fighting during the Chinese Civil War and was established in September 1950. Throughout the 1950s and early 1960s the Soviet Union provided assistance to the PLAN in the form of naval advisers and export of equipment and technology. Until the late 1980s, the PLAN was a large reverence and littoral force (brown-water navy). However, by the 1990s, following the fall of the Soviet Union and a shift towards a more forward-oriented foreign and security policy, the leaders of the Chinese military were freed from worrying over land border disputes, and instead turned their attention towards the seas. This led to the development of the People's Liberation Army Navy into a green-water navy by 2009. Before the 1990s the PLAN had traditionally played a subordinate role to the People's Liberation Army Ground Force. Chinese strategists term the development of the PLAN from a green-water navy into "a regional blue-water defensive and offensive navy."

With personnel strength of 255,000 servicemen and women, including 10,000 marines and 26,000 naval air force personnel, it is the second largest navy in the world in terms of tonnage, behind only the United States Navy, and has the largest number of the major combatants of any navy. The PLAN traces its lineage to units of the Republic of China Navy who defected to the People's Liberation Army towards the end of the Chinese Civil War. In 1949, Mao Zedong asserted that "to oppose imperialist aggression, we must build a powerful navy'. The Naval Academy was set up in Dalian on 22 November 1949, mostly with Soviet instructors. The navy was established in September 1950 by consolidating regional naval forces. It then consisted of a motley collection of ships and boats. The Naval Air Force was added two years later. By 1954 an estimated 2,500 Soviet naval advisers were in China, possibly one adviser to every thirty Chinese naval personnel, and the Soviet Union began providing modern ships. With Soviet assistance, the navy reorganized in 1954 and 1955 into the North Sea Fleet, East Sea Fleet, and South Sea Fleet. In shipbuilding, the Soviets first assisted the Chinese, then the Chinese copied Soviet designs without assistance, and finally the Chinese produced vessels of their own design.

Eventually Soviet assistance progressed to the point that a joint Sino-Soviet Pacific Ocean fleet was under discussion. In the 1970s, when approximately 20 percent of the defense budget was allocated to naval forces, the Navy grew dramatically. The conventional submarine force increased from 35 to 100 boats. The Navy also began development of nuclear attack submarines (SSN) and nuclear-powered ballistic missile submarines (SSBN).

The first class of submarine to be operated by the PLAN was the Soviet Whiskey-class. The Whiskey was imported from the Soviet Union and subsequently built in considerable numbers and served until the last few remaining boats were removed during the mid-1990s. The second type to be operated also owed its origins to the Soviet Union, the Romeo-class, they were built under license in China as the Type 033 submarine. Production of this submarine took place from the late 1960s until the late 1980s. The Type 033 went on to form the backbone of the PLAN submarine forces, and has been estimated that more than 100 may have been produced for the PLAN and for export. In 1994 it ordered two Kilo-class 877EKM type submarines from Russia, which were delivered by 1995. In 1996, two improved Kilo-class 636 submarines were ordered, delivered between 1997 and 1998. By the late 1990s, a large number of 033s had been retired from active duty and pulled into reserves. A handful of upgraded hulls remained in service until the late 2000s for training and other limited purposes. In 2002, a $2 billion deal was signed for eight more Kilo-class 636, these submarines particularly fitted with the capability of launching the Russian Novator 3M-54E Klub S cruise missile capable of engaging land and sea targets at 220 km. The Kilo class represents a huge leap forward in the PLAN submarine fleet. Originally a Soviet design in the 1980s, the Kilo-class was meant to be one of the world's quietest class of submarines. With 12 Kilos operational by 2006, it is unclear whether PLAN will buy more of these potent vessels. Despite the purchase of the Kilo-class, the PLAN has continued to develop indigenous designs.

Nuclear submarines have been envisaged in the PLAN since the 1950s. Despite ambition and a long history of development, the acquisition of nuclear submarines has been a difficult process. The Cultural Revolution greatly disrupted nuclear submarine development. The Sino-Soviet split prevented any Soviet assistance in nuclear propulsion, and these propulsion problems have been troublesome to this day.

The first Chinese nuclear-powered submarine was laid down in 1967 but not completed until 1974, the Type 091 submarine (Han-class). China became the first Asian country and fifth nation in the world after US, USSR, UK and France to build a nuclear propelled submarine. The Han-class experienced more than 20 years of development, with the last of the class not being commissioned until 1990. Since their commission the class has gone through major upgrades and numerous refits with the remaining boats having been greatly refitted with new sonars and anechoic tiles (which reduce noise levels). The Han-class has mostly operated in local waters, but since the 1990s, they have been used more aggressively. A Han-class shadowed a US carrier battle group in the mid-1990s, and more recently, operated around Japanese waters, prompting a Japanese task force to chase the submarine out of its territory. The Han-class submarines are far from being as capable or effective as their American Los Angeles-class counterparts, but with recent improvements they can pose a great threat by operating deep in the Western Pacific and attacking targets that are less well-protected. A new class of submarine has been in development since the 1980s, when the PLAN first sought a replacement for the Han-class. Little information has emerged about the Type 093 submarine (Shang--class), but it is believed to have some Russian influence. The 093 design maybe comparable to the Russian Victor III-class, signifying a significant step forward for Chinese nuclear attack submarines. The 093 has been the focus of much attention from US and Asian military analysts. Its improved capabilities will undoubtedly increase PLAN power in the region and its ability to carry war to the West Pacific. Such submarines can escort future Ballistic missile submarines as well as attacking US Navy carrier battle groups in the deep ocean. Two Type 093 submarines are in service with four more on the verge of being commissioned. China is constructing a major underground nuclear submarine base near Sanya, Hainan. The Daily Telegraph on 1 May 2008 reported that tunnels were being built into hillsides which could be capable of hiding up to 20 nuclear submarines.

In September 2015, satellite images showed that China may have started constructing its first indigenous carrier. At the time, the layout suggested a hull to have a length of about 240 m and a beam of about 35 m. The incomplete bow suggests a length of at least 270 m for the completed hull.

Japan has raised concerns about the PLAN's growing capability and the lack of transparency as its naval strength keeps on expanding. China has reportedly entered into service the world's first anti-ship ballistic missile called DF-21D. The potential threat from the DF-21D against U.S. aircraft carriers has reportedly caused major changes in U.S. strategy.

China has declared 'No First Use' which translates that she will not use nuclear weapons on any country. This statement always has an unwritten and hidden rider 'if any country uses a nuclear weapon on China, she has the capability of the Second Strike' from a nuclear powered submarine equipped with nuclear missile which she would use. It is believed that China has one of the world's most secretive nuclear weapons programs. This doctrine of 'No First Use' is not accepted as a sincere description of China's attitude toward its ever-increasing nuclear arsenal. When China had gained entry to the NSG, it was President Bush who overruled his Republican and Democrat colleagues having a 'deep distrust' of China and were of the opinion that China is one of the world's "principal sinners" when it comes to proliferation. Thus the fundamental question is why the world community should believe China's promise of 'No First Use', as there is a doubt on the integrity. It is a common knowledge that nuclear missiles are not kept 'ready to fire' at a specified target by merely pressing a button; even the warheads are not normally kept mated with missiles. It would take quite some time to launch 'The Second Strike' after the orders are given from the top, may be after some consultation. Submarines are almost impossible to be detected, thus Ballistic missile submarines can be used either first or second-strike weapons; as matter of fact submarines can be used to launch a first strike without any warning. These missiles housed in submarines are not likely to be destroyed like the land-based missiles during any strike by the enemy. Whether the submarine based missiles are designed to fulfil the first strike or the second strike role, the target data will have to be programmed/stored on board the submarines. One can argue that at present, the submarine launched missiles are probably the most efficient way of deploying nuclear weapons for any country wishing to have a credible deterrent. All of the five legally recognized NWS as per NPT, except China, possess and deploy nuclear-armed submarines, and some other states are putting their best effort to build/acquire this nuclear deterrent. Thus China, a late developer in this game, will certainly add more submarines with nuclear missiles and join USA, USSR, UK, and France whether these four nations like it or not. As a late developer, in comparison with the other four legally recognized nuclear weapon states, China will rely heavily on ballistic missile submarines to provide a large proportion of its nuclear deterrent China knows that survivability is measured on perception and ballistic missile submarines are the best insurance policy against unknown future threats.

Nuclear submarine deployment not only deters a first strike, also prevents a conventional conflict from escalating into a nuclear exchange. This is the reason why other countries too would like to build/acquire the nuclear submarines so that in case of a war any nuclear strike from the enemy, in retaliation, is prevented; though this option is very expensive but becomes a matter of prestige for any country. Thus hopefully, because of the threat of mutually assured destruction, the third world war may not take place. There is a quote fromLord Louis Mountbatten 'If the Third World War is fought with nuclear weapons, the fourth will be fought with bows and arrows.'

The news item in CHINADAILY in April 2016 says that China is planning to build about 20 floating reactor platforms to meet the demand of maritime atomic propulsion, driving a market that could be worth tens of billions of dollars, a report said on Thursday. The country's first-ever floating nuclear plant is about to start the final assembly in Huludao, a coastal city in Liaoning province, and it will be built by Bohai Shipbuilding Heavy Industry Co Ltd, a unit of China Shipbuilding Industry Corp, according to 'e world ship', a Shanghai-based maritime industry information provider." CSIC is the first company with the permission to construct the floating nuclear-powered vessel, and it aims to become the strongest builder of floating nuclear platforms within five years," the report quoted Wu Zhong, general manager of CSIC Asset Management Co Ltd, as saying during an expert review held this week. Zhu Hanchao, deputy chief engineer of the CSIC 719 Research Institute, said that the average cost of a pilot project is about 3 billion yuan ($463 million), but it will be able to generate sales of 22.6 billion yuan in 40 years, the life span of the vessel. He said that the cost could go lower as the company realizes mass production of such platforms. At the end of last year, the shipbuilder got the nod from the National Development and Reform Commission, the country's top economic regulator, to start research for a demonstration project of the platform. Earlier this year, China General Nuclear Power Group signed a strategic cooperation agreement with CSIC to develop a reactor design—200-megawatt ACPR50S for the offshore nuclear power platform. CGN is currently working on the preliminary design for ACPR50S, which is expected to start construction in 2017 and be commissioned by 2020. The floating nuclear power plant, which can be equipped inside a section of the vessel, is often used to supply stable electricity not only to remote areas, but also to large industrial facilities such as seawater desalination plants and offshore oilfield exploration rigs, CGN said." The project has a wide range of civilian applications in providing safe and stable energy for maritime resources exploration and development," CGN said.

NPT, NWS, NNWS, NSG, MTCR

Ever since the atomic Bombs were dropped in Hiroshima and Nagasaki, nuclear weapons have become synonyms with great power status and prestige and a source of global influence. The US being the first, was keen to maintain its monopoly but others followed suit. USSR in1949, UK in 1952, France in 1960 and China in 1964 tested the nuclear weapons. They joined together, US, USSR, U K, France and had no choice but to include China whether they liked or not, in preventing others to have nuclear weapons, ignoring the fact that the very existence of Nuclear Weapons with associated 'aura' is an incentive for others to get it. This led to 'asymmetry' of 'haves' and 'have not's'. All other countries had to work in the nuclear energy field, whether peaceful or otherwise, in this 'technology denial regime'.

The Treaty on the Non-Proliferation of Nuclear Weapons, commonly known as the **Non-Proliferation Treaty or NPT**, is an international treaty whose objective is to prevent the spread of nuclear weapons and weapons technology, but to promote cooperation in the peaceful uses of nuclear energy, and to further the goal of achieving nuclear disarmament and general and complete disarmament.

Opened for signature in 1968, the Treaty entered into force in 1970. On 11 May 1995, the Treaty was extended indefinitely. Five states are recognized by the NPT as **nuclear weapon states (NWS)**, because they tested the Nuclear Weapons before 1970 when the treaty entered into force all other states are **NNWS, Non Nuclear Weapons States**. These five nations are also the five permanent members of the United Nations Security Council. India argues that the NPT creates a club of "nuclear haves" and a larger group of "nuclear have-nots" by restricting the legal possession of nuclear weapons to those states that tested them before 1967, but the treaty never explains on what ethical grounds such a distinction is valid. The former Chairman of the Indian Atomic Energy Dr. Chidambaram who headed the nuclear weapons programme always remarked about the absurdity of the NPT by citing an example "Just because one does a graduation after 1970, his/her degree is not recognized though he/she followed the same syllabus and passed the same examination". India's then External Affairs Minister Pranab Mukherjee said during a visit to Tokyo in 2007: "If India did not sign the NPT, it is not because of its lack of commitment for non-proliferation, but because we consider NPT as a flawed treaty and it did not recognize the need for universal, non-discriminatory verification and treatment." . India has always supported universal, non-discriminatory arrangements. Precisely for this reason she is not a party to NPT. India's concern for proliferation of nuclear weapons is second to none. As a responsible State, India have stringent export control regimes. Physical protection measures are in place in all her nuclear installations India has State System of Accounting and Control (SSAC) in place. On January 22, 1965, India's Plutonium Plant was inaugurated at Trombay by the then Prime Minister Lal Bahadur Shastri. And India tested her first nuclear device in 1974, (The terminology used is PNE, 'peaceful underground nuclear experiment') becoming the sixth nation to test the nuclear weapon but not included in the list of NWS. India tested a number of nuclear weapons in May 1998 and declared 'No First Use'. India was among the few countries to have a no first use policy, a pledge not to use nuclear weapons unless first attacked by an adversary using nuclear weapons, however India's NSA (National Security Advisor) Shivshankar Menon signaled a significant shift from "no first use" to "no first use against non-nuclear weapon states" in a speech on the occasion of Golden Jubilee celebrations of the National Defense College in New Delhi on 21 October 2010, a doctrine Menon said reflected India's "strategic culture, with its emphasis on minimal deterrence". This statement becomes ambiguous because of the words 'non-nuclear weapon states' because of the major controversy about NPT which recognizes only 5 NWS, the remaining are NNWS! In the opinion of the author, the better words would be 'the states not possessing nuclear weapons.'

More countries have adhered to the NPT than any other arms limitation and disarmament agreement, a testament to the Treaty's significance. Most of the countries were coerced and pressurized by the five NWS to sign the treaty. The NPT is often seen to be based on a central bargain: "the NPT non-nuclear-weapon states agree never to acquire nuclear weapons and the NPT nuclear-weapon states in exchange agree to share the benefits of peaceful nuclear technology and to pursue nuclear disarmament aimed at the ultimate elimination of their nuclear arsenals.

The 1978 Nuclear Non-Proliferation Act (NNPA) of U.S sought permit to supply of nuclear related exports to NNWS, only if such states accept IAEA comprehensive Safeguard Agreement, not to establish any new enrichment or reprocessing facilities. The existing agreement for cooperation was made subject to review on a case by case basis. In the 90s, the safeguards were further strengthened because of Iraq and DPRK. A protocol, additional to a safeguard agreement was introduced to detect undeclared nuclear material and activities including Environmental Sampling and remote monitoring.

Among the five states recognized by NPT as nuclear weapon states, China and France signed the NPT in 1992, while the Soviet Union (now Russian Federation), the United Kingdom and the United States had signed in 1968. A total of 191 states have joined the Treaty. North Korea ratified the treaty on 12 December 1985, but gave notice of withdrawal from the treaty on 10 January 2003 following U.S. allegations that it had started an illegal enriched uranium weapons program, and the U.S. subsequently stopped the fuel oil shipment. Three other states, India, Pakistan and North Korea have openly tested and declared that they possess nuclear weapons, while Israel has had a policy of opacity regarding its nuclear weapons program. On 18 September 2009 the General Conference of the International Atomic Energy Agency called on Israel to open its nuclear facilities to IAEA inspection and adhere to the non-proliferation treaty as part of a resolution on "Israeli nuclear capabilities". The chief Israeli delegate stated that "Israel will not co-operate in any matter with this resolution. Four UN member states who never joined the NPT are India, Israel, Pakistan and South Sudan.

After developing a number of sophisticated nuclear weapons and the associated delivery systems, all these 5 designated NWS, went on to develop/acquire nuclear submarines to be able to launch nuclear tipped missiles while submerged deep underwater and at the same time not being 'visible' to the enemy. Thus they all developed 'Second Strike' capability as well.

Nuclear Suppliers Group (NSG) is a multinational body concerned with reducing nuclear proliferation by controlling the export and re-transfer of materials that may be applicable to nuclear weapon development and by improving safeguards and protection on existing materials. The NSG was founded in response to the Indian nuclear test in May 1974 and first met in November 1975. The test demonstrated that certain non-weapons specific nuclear technology could be readily turned to weapons development. Nations already signatories of the Nuclear Non-Proliferation Treaty (NPT) saw the need to further limit the export of nuclear equipment, materials or technology. Another benefit was that non-NPT and the nations not in the list of-Zangger Committee, also known as nuclear exports committee, then specifically France, could be brought in. As of 2014 the NSG has 48 members. China has opposed India's bid to get Nuclear Suppliers Group membership on the ground that it was yet to sign the Non-Proliferation Treaty. "Apart from India, other countries expressed their willingness to join; China's position is not directed against any specific country but applies to all the non-NPT members"

A few days back in June 2016, the Prime Minister Modi made a 5 nation tour; he built up the support for NSG membership with Switzerland, USA and Mexico. He also addressed the US congress and got standing ovation. US has expressed strong support for India's entry into the 48 member Nuclear Suppliers Group; but China remains adamant in its stance in opposing India's entry into the NSG so the issue is now the latest battleground in the growing Sino-Indian affairs. Beijing is making it clear that it intends to make life difficult for India.

China herself had gained membership to the NSG in May 2004, despite the opposition by several US lawmakers - both Republicans and Democrats - who were overruled by US President George W Bush. One Republican lawmaker called China one of the world's "principal sinners" when it comes to proliferation, and a Democrat said he had a "deep distrust" of China.

The **Missile Technology Control Regime (MTCR)** was established in April 1987 by the G7 countries: Canada, France, Germany, Italy, Japan, Great Britain, and the United States of America. The MTCR was created in order to curb the spread of unmanned delivery systems for nuclear weapons, specifically delivery systems that could carry a payload of 500 kg for a distance of 300 km. At the annual meeting in Oslo on 29 June - 2 July 1992, it was agreed to expand the scope of the MTCR to include nonproliferation of unmanned aerial vehicles (UAVs) for all weapons of mass destruction. Prohibited materials are divided into two Categories, which are outlined in the MTCR one is Equipment and the other is Software, and Technology. Membership has grown to 35 nations. India joined on 27th of June 2016 adhering to the MTCR Guidelines unilaterally. In 2004 China applied to join the MTCR, but members did not offer China membership because of concerns about China's export control standards. China desperately wants into the MTCR and it is very likely that she opposed India's NSG bid for that reason - a quid pro quo. "Allow us into the MTCR and we will not oppose India's entry into the NSG", might well have been what China was whispering into ears that mattered.

NUCLEAR PROPULSION IN INDIA

India too went through somewhat similar situation like the US and faced delays in starting the program of nuclear propulsion; the delays were not entirely due to a difference of opinion and the lack of confidence between the Navy and the Atomic Energy; it had taken a long time for India to graduate for tackling this advanced technology. The published US experience probably became the role model, with positive and negative aspects for the others to follow. It is interesting to take look at the Indian Navy's history and the background after independence, prior to India starting its program for nuclear propulsion for the submarines, in spite of the fact that this was a period of 'nuclear technology denial regime'.

Indian Navy and the Submarines

The Indian Navy continued to grow, after independence in 1947, only by adding surface ships. This was commonwealth's naval strategy, with emphasis on anti- submarine warfare against the soviet submarine threat in the Indian Ocean. India's ever dependence on Britain on all naval acquisition defined the Indian navy's force structure. The submarine arm was considered offensive in nature; the Royal Navy officers commanded the Indian Navy till 1958 when Vice Admiral Katari was appointed the first Chief of Naval Staff. The British naval officers were of the view that the submarines were far too sophisticated for the Indian Naval personnel to operate; it was too early for them to venture into the field of submarines, torpedoes and mines. The defeat from China in 1962 brought a change in India's attitude. The Indian Ocean region was witnessing a vacuum and rapid development of Indonesian and Pakistan navies were of great concern, after the withdrawal of British from the Indian Ocean. In 1964, the US administration agreed to a long term military assistance to India to counter China; the same year China tested the nuclear weapon. Though Pakistan had acquired a submarine from the US in 1963, India was denied by the US as well as Britain; India started seeking Soviet assistance for its defense requirements. Soviet Union expressed interest in building India's submarine arm.

The 1965 war with Pakistan revealed the Indian Navy's limited capabilities. In September 1965, Soviet Union agreed to sell 4 Foxtrot class submarines to India. By the middle of 1966, Indian sailors arrived at the Soviet Naval base for training. The first submarine joined the Indian fleet in 1968 thus initiating the long desired submarine arm in the Indian navy and this also opened up the possibility of long term naval cooperation and arms transfer.

The Beginning-Theoretical Studies

Induction of the first submarine also had major implications for the Indian Navy's decision to initiate a nuclear submarine program. Some scholars have suggested that during this time, the subject of nuclear submarines for the Indian Navy was also broached, because in 1968, two engineers from the navy, Gurmeet Singh (He rose to the rank of Rear Admiral and spent most of his tenure in the nuclear propulsion programme) and V K Chadha, were deputed to join the 'postgraduate course in nuclear science and technology' along with the trainees of Atomic Energy in the training school of BARC, Bhabha Atomic Research Center in Trombay and the feasibility studies were undertaken for nuclear propulsion by a small team of scientist from BARC and engineers from the navy, stationed in BARC. In 1970, NPT had come into force; it became evident that in future some help from other countries may continue to be available for naval vessels but India would have to design and build the nuclear propulsion plant all by itself. Being discriminatory, India had not signed the NPT.

The spadework for the induction of nuclear propulsion was commenced long before the date normally ascribed to it. Not sufficient credit is given to the foresight of the naval leadership. Since the early 1950's, design engineering was part of the advanced engineering courses undergone by selected Indian Naval engineer officers at the Royal Naval Engineering College at Plymouth and Cranfield institute of Technology in the UK. Similarly, the naval architects obtained their post-graduation at the Royal Naval College at Bath, Greenwich in UK and the Grechko Naval Academy at Leningrad in the USSR. Ship, submarine and its associated machinery design formed a part of their studies. Additionally, in the 1980's, as part of the HDW s/m contract, a multidisciplinary design team was also deputed to Professor Gabler's submarine Design institute at Kiel and Lubeck in Germany. Some of these officers, who graduated from these institutions, later joined the 'nuclear submarine programme' and contributed a lot.

Evidence now shows that even Prime Minister (PM) Nehru never locked the door on keeping the option open for India to weaponise the nuclear genii under national security compulsions. This policy was sustained by most of the PM's who followed. Dr. Homi Bhabha, all along, had been advocating to PM Nehru that nuclear technology must be nurtured for such an eventuality; in the correspondence, between Nehru and Bhabha, one can conclude reading in between the lines. The real impetus to the project came after the Soviet Navy demonstrated to PM Indira Gandhi, the capability of the nuclear submarine, in deterring the attempts of the US navy to intervene in the Bay of Bengal, during the 1971 war. In February 1972, Grechko, the soviet Defense Minister, is reported to have brought to the notice of Ambassador D.P Dhar and General Manekshaw how a soviet nuclear submarine stalked the US Navy's nuclear powered carrier USS Enterprise in December 1971. Thereafter, Indira Gandhi gave unstinted support to the project. As a matter of fact, she became the driving force behind it. In this context, the suggestion that the Soviets offered us the nuclear submarine assistance, even before the indo-soviet strategic Partnership agreement was signed and Germany conceded to signing NPT in 1970, is inconceivable. In 1974, India conducted her PNE (Peaceful Nuclear Explosion) ; international community, and the US in particular, were very sore and the regime of 'denial' to India started. India never signed NPT on the ground that it is 'discriminatory'. The US declared that it will not supply enriched uranium for the Tarapur Reactors; BARC started studying the possibility of using Plutonium instead of enriched uranium to run the reactors; a number of irradiation experiments were also undertaken and a few assemblies were fabricated with the mixed oxide fuel and successfully irradiated in one of the Tarapur reactors. This became the starting point of the study about the possible use of mixed oxide fuel in a future propulsion reactor; in addition, mixed carbide fuel with 70 % plutonium was already chosen for the fast breeder test reactor at Kalpakkam.

In 1976, a team from Navy, initially under the leadership of Captain P N Agarwal and later Captain B Bhushan, (He later became Vice Admiral and headed the ATV after V Admrai M K Roy, for about 10 years) was posted in BARC under the overall guidance of Dr. P R Dastidar, Director Reactor Group, BARC, to get trained and study various design options with help from BARC. Code named PRP, (Plutonium Recycling Project-land based prototype propulsion plant), a number of scientists and engineers from the Reactor Group, BARC, mainly from Reactor Engineering headed by S K Mehta, Reactor Control headed by S N Seshadri, Reactor Physics headed by B P Rastogi and Reactor Feasibility studies headed by L G K Murthy, participated in evaluating various reactor systems and design options. Broad parameters for the nuclear propulsion plant were provided by the naval team, based on the continuously updated information from the naval headquarters. This naval team also used to work out the parameters of the non-nuclear systems and the overall configuration, for the future submarines in mind. The few published literature/ unclassified reports from the US, UK, France and the Soviet Union also formed the basis. Information from engineer officers and naval architects, who had done the post-graduation in UK and USSR, used to be quoted in the discussion. Some R&D was done including fabricating scale down models in Hall No 7 in BARC. Several project reports were rendered by the joint BARC-cum-naval teams code named 932. These reports were hotly debated, at times leading to unpleasantness, even to the extent of undermining each other's capability; in particular the BARC members were not ready to give up the choice of mixed oxide fuel till they find out the scientific reason why it could not be used in the compact reactor system; by that time the R & D for enriching the uranium was in full swing but the actual plant in Mysore for enriching the uranium was far off. This however did not come in the way of designing other components or working out the layout, the fuel in a reactor is replaced periodically and a reload could always be different in design or the material composition. Later the enrichment plant was set up and the capacity was enhanced when we concluded that due to Reactor Physics considerations in a compact core, we must use enriched uranium fuel.

For almost 7 years, both teams interacted with each other; this was not easy as the naval culture is autocratic while BARC culture is very democratic; in the beginning, the naval officers had the feeling that they were in BARC to get the work done from BARC scientists and engineers, like the way they handle the personnel from public and private organizations, on the other hand, BARC scientists and engineers felt that the naval officers are unnecessary interfering without having required technical knowledge. The things eased out after a while, the friendships developed and both sides started meeting socially over a drink in the evening. By this time, it was also recognized by all that both Navy and Atomic Energy have to find & evolve a common Administrative & Management structure, if the project had to take shape. Dr. Ramanna who had left for Delhi in 1978 to head the DRDO returned in 1981 and re assumed the office of Director BARC; with his experience in the defense ministry and the closeness to the seat of power in Delhi, he started focusing more on the nuclear propulsion; on hind sight, one can conclude that a lot of spade work must have been done. In 1983 a final project report was generated for discussion and approval, indicating the broad parameters of the submarine like displacement, design depth and the overall dimensions of the submarine. The knowledge of Soviet Foxtrot submarines and the operating experience, also formed the basis of the report; in addition some experience of the single hull submarines 'Shishumar Class' being built at Mazagon Docks on 'technology transfer' from German yard Howaldtswerke-Deutsche Werft (HDW) under the internal classification Type 1500, was also available.

Admiral Revi writes—"*During the same time between 1981 and 1983, discussions were held between India and the USSR about the possibility of technical help from USSR for the nuclear propulsion programme. In April 1981, during chief of General staff, the Marshal of the Soviet Union, Ogarkov's visit to India, he formally enunciated the so called Ogarkov Principle that laid down the parameters of the assistance feasible. Their insistence on compliance with the safeguard clause prior to any agreement on transfer of technology (tot) for the reactors was not a surprise but was not acceptable to India. In July 1981, Dr. Raja Ramanna's delegation was taken to Murmansk and shown a nuclear propelled submarine. After extensive negotiations, the agreement was concluded in April 1982, conforming to the Nuclear Proliferation Treaty (NPT) and Missile Technology Control Regime (MTCR). All subsequent Soviet/Russian cooperation on the nuclear submarine programme flows out of that umbrella agreement. The issue of detail design assistance on the reactor was left to be considered separately. For gaining operation and maintenance experience, the Soviet Union would lease a nuclear propelled submarine to the Indian Navy before the induction of its own nuclear propelled submarine. Physical access to the submarine and its propulsion plant to the designers would be permitted to familiarize them with the layout and space constraints. To begin with, two sets of crew of the Indian Navy would be trained for operations and maintenance and the assistance would be provided for setting up the submarine building infrastructure. Finally, the K-43 Charlie Class, christened INS Chakra was leased to India on 1 September 1987 and reached its base in Vizag on 3 February 1988 after a long journey. The Soviets said that the submarine was transferred for helping train the Indian Navy in operating nuclear submarines. During its service with India, it was partially manned by a Soviet crew, who reportedly did not allow Indians into the missile room and into the reactor compartment and this is believed to be a reason for the termination of the contract after 3 years. The lease of Chakra reportedly helped India gain first-hand experience in handling a nuclear submarine that helped evolving the layout for the Arihant-class of submarines."*

In 1983, the naval team located in BARC under the leadership of Captain B Bhushan took upon the challenge of familiarizing the crew, chosen to go to Soviet Union, about the nuclear propulsion. The crew included the first Commanding Officer of Chakra Captain Ravi Ganesh who became a Vice Admiral and later took over from Admiral B Bhushan as Director General, Advanced Technology Vessel (ATV). This exercise was conducted at the naval training center. The scientists and the nuclear engineers from BARC participated in this exercise to teach Reactor Physics, Reactor Engineering, Reactor Control, Shielding Design and other nuclear aspects.

At the same time, a classified BARC office order was issued appointing Anil K Anand of Reactor Engineering Division, BARC as the Project Manager (PM) of PRP to be built in the DAE complex at Kalappakam. The Director BARC at that time was Dr. Ramanna and he was also scheduled to take over as the Secretary DAE/Chairman AEC in a few months' time after the superannuation of Dr. H N Sethna. At the same time, there was a talk that the C-in-C East, Vice Admiral M K Roy, after his retirement in a few months, would be reappointed by the Government as Director General of ATV programme, to design and build indigenously, the nuclear propelled submarines. Anand, the designated PM, PRP from BARC accompanied Captain Bhushan and other naval officers to the city of Visakhapatnam (nicknamed Vizag) to meet Admiral Roy; the other 2 important members of the team were Commanders Gurmeet Singh and R S Chaudhry; later, both rose to become Rear Admirals. Some preliminary discussion was held in a brief meeting and modalities were discussed. Later, after the Government orders were issued, Anand was again called by Admiral Roy to Vizag to discuss about the future organization structure of ATV. The ATV project would be almost entirely managed by naval personnel, though under the DRDO umbrella. Admiral Roy was very clear about the location of ATV headquarters at Delhi, in Sena Bhavan but he was not sure if the PM PRP would be better relocated at Delhi or continues in BARC. After discussing the pros and cons, it was decided that Anand should continue in BARC Trombay as the PRP, after a few years, would need hundreds of engineers and scientist in BARC to work on the project and it would be an impossible task to coordinate remotely. It would be better to decentralise the design and procurement activities not only in case of propulsion plant but also in case of other work centers in future that would be headed by independent chiefs. All would attend the 'Project Management Board Meetings' regularly in ATV Headquarters. This was the most challenging and important programme undertaken by the Atomic Energy and the Indian Navy; the other two which were mentioned in the corridors were the 'Main Battle Tank' for the army and 'Light Combat Aircraft' for the air force.

ATV Organization Setup

The naval team thus was wound up from BARC as the time had come to move away from the paper design, start the detailing of the program and to design and build the systems and facilities. The new management structure was evolved to design and build the land based prototype propulsion plant and to design and build the future nuclear submarines. The credit for this enormous task goes to 1) Dr. Raja Ramanna, Director BARC and the Chairman designate of the Atomic Energy; 2) Dr. V S Arunachalam, Director General for Defense R & D and the Scientific Advisor (SA) to the Defense Minister; and 3) VAdm M K Roy C-in-C East and the Director General designate of the ATV (Advanced Technology Vessel) program, after retirement. All three were very good friends ; Dr. Ramanna knew DRDO and the working in the Defense Ministry very well as he had been the DRDO chief earlier for a few years. Dr. Arunachalam had also spent his formative years in Atomic Energy, he is from the second batch of BARC training school, a brilliant metallurgist; after spending a few years in BARC, he was transferred to DRDO and quickly rose to become the DRDO chief. ATV organization was to look after all the non-nuclear engineering systems in the submarine design and build the required facilities for the same. The naval team which was wound up from BARC became the part of ATV. The ATV Apex Board was chaired by the RM (Raksha Mantri/Defense Minister) and members consisted of SA to RM, CNS, Secretary DAE, Secretary Finance and DG ATV; at the next level was the Technology Management Board – Chaired by SA to RM; both these boards met occasionally. The third level was the PMB, the Programme/Project Management Board chaired by the DG ATV and attended by heads of all Work Centers and the FA, the Financial Advisor ATV (he was also the FA Defense R & D) met regularly, at least once a month. Defense R & D was to coordinate all the activities of the program including those of the defense laboratories and chair the finance review meetings regularly. The ATV was managed predominantly by the serving naval personnel seconded by the navy but there was an unwritten rule that they would not wear uniforms and exhibit their ranks unless attending a formal naval function. They had to interact with a large number of scientists and engineers from the DRDO, the Atomic Energy and a number of public and private enterprises where the setup is democratic and the discussions on technical matters are held in an open manner among juniors and seniors.

By this time, Dr. Ramanna had taken over as the Chairman Atomic Energy and Dr. P K Iyengar had succeeded as Director BARC. A site for building the land based prototype propulsion plant was chosen at the sea shore in the DAE Complex at Kalpakkam, in-between the Madras Atomic Power Station and the IGCAR (Indira Gandhi Center for Atomic Research) where the fast breeder test reactor was recently built. In the BARC, a new division RPD 'Reactor Projects Division' was carved out to carry out the design of the PRP, the land based nuclear propulsion unit, Anand, the PM, PRP was appointed as the head of this division. RPD included all the engineers from the erstwhile 'Fuel Design and Development Section' headed by Anand, this section included among others, Basu (Present Chairman of AEC) and Vyas (Present Director BARC); some more engineers were transferred from other sections, totaling about a dozen in RPD. Basu and Vyas concentrated on the core and fuel design; Mehra, Yadav and Mukherjee on the Reactor Pressure Vessel, Steam Generator, Pressuriser and the 'Support Structure' for NSSS (Nuclear Steam Supply System), Agarwal was the shielding expert, Grover did Thermal Hydraulics calculations and Nangia, D P Rao, P Sreenivas and Ramaswamy concentrated on the process design of the auxiliary equipment and reactor compartment layout; later after finishing his assignment in Dhruva, Utge also joined this group. Reactor Physics support was provided by Dr. Dwivedi and his group and the Reactor Control and Instrument support came from the head of Reactor Control Division, Govindarajan along with his two groups headed by G P Srivastava and R K Patil. A few months later, 8 engineers (AC Bagchi, Anupam Sharma, Devesh Kumar, T Narayanan, AK Sharma, Mrs. Salvi Vasan, PR Patil and Shri Sujay Bhattacharya) joined RPD in August 1984 after passing out the post graduate course from BARC Training School; these engineers were sent for training in the only LWRs (Light Water Reactor) at Tarapur. 8 more engineers (KV Ravi, MM Rajput, KG Prabhakaran TS Srinivasan, P Goverdhan, Mrs. PM Geetha, ML Yadava and S Venkataraman) joined in August 1985 and were also trained at Tarapur; thus the strength of RPD more than doubled in about a year. It seems, we more or less replicated the administrative &management structure which included Navy, DRDO, and DAE personnel, similar to the one chosen by the US in late 40's, to build the project; the difference being that, by this time, we had designed and built not only the power reactors for electricity generation but a fast breeder test reactor as well. The industrial base for the nuclear technology had also been successfully established within the public sector and the private sector. The major difference, in propulsion reactor, as compared to the power reactors, is the requirement of much higher power per unit of space & weight and not able to rely on gravity for shutting down the reactor; in addition operating in rough seas, requires the plant to be designed to withstand vibrations, large vertical & horizontal forces, rolling and pitching. The rate of power rise is an order of magnitude higher than the land based power reactors, so that the submarine can escape fast if in danger. The refueling interval is comparatively very large, requiring higher enrichment of U235 and gradual dissipation of burnable poison.

Independent 'Design Safety Review Committee'

There is a very good tradition in the BARC of constituting an independent 'Design Safety Review Committee' (DSRC) for every project when the design starts. The designers present their design to this committee which reviews the design with a view to assess if all the safety aspects have been covered. The members, from various disciplines, are equally good in their respective fields and working on their own projects. The Design Safety Committee was set up under the chairmanship of Dr. M Srinivasan a very competent Reactor Physicist who had worked on Plutonium and Uranium 233 cores and headed the Neutron Physics Division in BARC; at that initial stage, more emphasis is on the reactor physics and fuel design which go through a lot if iteration. After a few years, when that phase was over, Dr. Srinivasan requested the Director BARC to reconstitute the committee with a good mechanical engineer as the Chairman. Thus the committee was reconstituted with Nanjundeswaran as the chairman; he was Director LWR (Light Water Reactor) Group in NPCIL (Nuclear Power Corporation and had a long experience in Tarapur right from the construction on wards up to its operation. After Nanjundeswaran's retirement, V K Chaturvedi from NPCIL took over as the chairman as he was looking after the Kudankulam project with 2, 1000 Mw VVERs. When Chaturvedi was designated as the CMD of NPCIL, he handed over the 'baton' to his next in command S K Jain. When the commissioning started and the O&M manuals started getting prepared, a new Chairman S K Sharma, the Director of Reactor Group at the BARC was appointed; rightly so as he was in-charge of all the operating reactors for research and isotope production, inside BARC.

Land Based Prototype

The first prototype with a number of engineering systems invariably is a small, 'scale down' model; after tackling the initial teething troubles and the successful operation, 'up scaling' is done and the required bigger and commercial version is designed. However in the case of propulsion plant for the submarine, due to a number of technical reasons, the first plant itself is of the same size and the specifications are the same, as required for installing on the submarine. The first unit is designed and fitted in the hull as a stationary land based unit which is later used for training the crews before they are sent to operate the submarines. The second unit, the sea going version, lags behind in design and fabrication; any changes or modifications in the land based unit are quickly incorporated in the sea going version. While the detailed design started in BARC Trombay, the civil works had to start at the same time at Kalpakkam. The first job was to design and erect the buildings and other civil structures to house all the systems of the propulsion plant, starting with the reactor, steam generators, turbine & condensers, turbo generators, propulsion shaft & the propeller and the dynamometer to absorb the generated power.

In the early 1985, 7 engineers, 3 from BARC—Anand , Nangia and Kamath and 4 from navy—Gurmeet Singh, Nadaph, Thukral and Kushwaha camped in the DAE guest house at Kalappakam for two weeks with a temporary office in the guest house itself to discuss, debate and work out the layout and the size of all the buildings and the infrastructure required and make the detailed report for execution. The local logistic support was rendered by the Chief Engineer, Civil V Krishnamurthy and the Director Engineering Services K Venkatraman of IGCAR; it was decided that they would be getting the civil works done for PRP. Keeping in mind the actual refueling requirements for the future submarines at a naval base and following the same guidelines, the refueling scheme was finalized earlier by Basu at RPD. This helped to work out the dimensions of the main plant building, designated No 1, to house the hull containing the propulsion plant. An upper limit of the hull diameter and the length of the truncated submarine required for installing the propulsion plant were assumed for freezing the dimensions of the building No 1 which was 100 meters x 20 meters. The crane capacity and the hook height was based on the requirements of the refueling scheme and the weight and overall dimensions of an individual component; finally a 100 ton crane at hook height of 23 meters was arrived at. Total of 17 civil structures including the water tanks were required; there was a huge sea water tank for the dynamometer. All the water, power and ventilation requirements for each building were worked out and the report got completed. This was the so called 'Red Report'. We used to work on this report from morning till late in the evening and after winding up for the day, before dinner & over a drink, we used to debate and argue on what had been accomplished during the day and chalk out the next day's plan. Thus there was no physical exercise; after first 2 days we started getting up at 5 in the morning and went for cycling from the guest house to Mahabalipuram and back, covering about 25 kilometers; fortunately there was a shop nearby who had 6 very old bicycles for hire, in any case, Kamath had dropped out of this cycling exercise due to some health reasons. When we presented the report to Admiral Roy, he appreciated the effort; we added that along with work, we have also passed the test of physical fitness by cycling 25 kilometers every morning; he smiled.

Before leaving Kalappakam, Gurmeet Singh and I called on C V Sundaram, the Director IGCAR, to thank him for all the help and the hospitality. We were in for a surprise, he gave us a long lecture on why this project should not be located in the DAE complex at Kalappakam, instead, any place can be selected on the long sea coast between Chennai and Vizag. He was a bachelor, teetotaler and a Gandhian workaholic; he said that the navy has a different work culture which would pollute his environment which has been built from scratch. He told us to convey his views to the authorities; while leaving we told him politely that he can also talk to Dr. Ramanna next time when he meets him.

The design work in RPD BARC was in full swing; about two months after submitting the project report prepared at Kalappakam, I got a call from the Chairman's office that Dr. Ramanna was traveling to Chennai next morning, on the way to Kalappakam and I should catch the same flight. At the airport in Mumbai, I was introduced to 2 other gentlemen who were accompanying him to Kalpakkam; one was Ravi Budhiraja (IAS Maharashtra Cadre on deputation, he retired as the Chairman Jawaharlal Nehru Port Trust) and the other Shivshankar Menon (On deputation from the Ministry of External affairs, he later became foreign secretary and the National Security Advisor) both were Deputy Secretaries in DAE. When we landed in Chennai, there were two cars to take us to Kalpakkam, one for the Chairman and the other for 3 of us; we traveled together but I could not talk about my work in BARC, I had learnt to be evasive, but at the same time kept up the conversation keeping in mind the principle of 'need to know' basis. The following morning Dr. Ramanna visited the site alone; I was already there with Krishnamurthy and Venkatraman, the site was leveled and cleared of debris etc. but no work had yet started. Dr Ramanna inquired as to why the civil works had not started yet; Krishnamurthy said that he is awaiting an office order along with the 'head of accounts' under which the expenditure would be booked. He was asked to get the required paper typed which the Chairman signed placing it on the bonnet of the car and the work started in full swing from the next day onward. All the expenditure on the project was compiled by IFA BARC which was periodically reimbursed by the FA ATV.

A similar situation was faced in the design office in BARC. We had started designing the Reactor Pressure Vessel, Steam Generators and Pressuriser and other reactor components. It was decided that BHEL, Tiruchy would establish facilities and manufacture this equipment; BHEL would also import material, specifications for which were generated jointly; they had been already working for quite some time and they wanted to regularize the procedures. K Balu, a friend of mine was the Director DPS (Purchase and Stores) of the DAE; I told him that we have to place an order on BHEL Tiruchy. He explained to me that Government rules do not permit a Purchase order to be placed even on a Public undertaking without calling for quotations. I reminded him that Dr. Bhabha, among other exemptions had also got an exemption on purchase of stores and materials without the intervention of the Director-General of Supplies and Disposals of Government of India. I tried my best to explain to him that this was for a classified project and I have the minutes of meetings in which the decision was taken; the meetings were attended by 3 Secretaries to the Government and CMD BHEL, but Balu explained that the rules in his DPS are much better than those in other government departments but still, it was not in his powers to place the order without any quotations and work quantification. We used to have a wonderful IFA in BARC, Shri Borkar. I explained the problem and the sensitive nature of the project; I also told him that establishing manufacturing facilities, procuring of the material etc. had to go in tandem with the design. There would possibly be mid-course corrections and some rework. Time and cost estimates could not be given but BHEL had promised that they would deploy all required resources to finish the job in the shortest possible time. I also requested him that if required he can talk to A K Mitra, the Financial advisor ATV, another wonderful gentleman with humane qualities. Shri Borkar himself prepared the draft, of the note for to me to write to the Director BARC, through the IFA. The matter was settled and the order was placed directly by BARC without involving DPS; BHEL would keep all the records of the expenses on this project, to be verified by IFA BARC, and get the expenditure reimbursed periodically.

The team of engineers in RPD used to meet regularly at Anand's house in the evenings to have social interaction and at the same time to continue the technical discussion and tackle administrative hurdles as there were a lot of uncertainties and question marks about the continuation of the project; it could be dropped any time along the way. One evening Anand was rather sad and quiet as he had received some disturbing information about the project which created a lot of uncertainty. The young engineers noted this and all of them started singing the famous song 'Hum hon ge kamyab ek din, hum ko hai vishwas'; (We will certainly succeed one day, we have the belief) this changed the environment and everyone enjoyed the evening there after. The young team had the determination. This kind of situation happened very often during the execution of the project. In addition, at times, there were discouraging remarks from other colleagues in BARC. M M Rajput once narrated an incident during his train journey from Kalappakam in 1987 "I happened to meet one of the senior BARC colleagues. During interaction after knowing that I am working for PRP he remarked that he knows this project; it is in the project stage since 1977 and it will never come up. My reply to the gentleman was 'Do you think I am a fool to work for the project and think that it will never be successful. The very fact that I am in the project and working for it, it will be successful'. I lived with this statement throughout my carrier keeping it in my heart. There were many occasion and difficult period for the project between 1995 and 2003 where it sounded like that particular gentleman was going to win but we overcame all the problems and difficulties and finally I won. Still today I live with same principle towards work I am involved."

During the first few review meetings of the ATV Programme Management Board in Delhi, the discussion primarily centered around the internal dimensions of the hull and weight; all the equipment designers of the propulsion plant, led by Anand for primary system and Gurmeet Singh for secondary system, demanded more space/area/weight whereas the naval architects and the hull designers led by R S Chaudhry had their own compulsion to restrict these parameters to the barest minimum. A number of iterations were gone through, before the upper limits were generated. The diameter of the hull and the cross section was first frozen and then the length for different compartments was worked out, based on the most compact possible layout. In some cases, by design, a component would have to be removed to have an access to another component needing maintenance.

The traffic through Chennai to Tiruchy and Kalpakkam increased; the logistics started becoming a problem and inconvenient. The journey

Chennai-Tiruchy-Chennai had to be performed by train as it was convenient even for those who were entitled to fly between Chennai-Delhi/Mumbai. ATV established a small office and an overnight transit accommodation at Adyar in Chennai; this was managed by a retired Commodore Krishnaswamy. This facility continued till the land was acquired at Pudupatnam and a separate residential colony was built about 8 kilometers from the PRP site. A team of engineers K G Prabhakaran and S Giridharan were posted at Tiruchy to follow up, the technical clarifications and inspection of the equipment. Walchandnagar, an overnight journey by train from Mumbai was the other place where a team of engineers P K Sharma and R K Mahajan were located. In addition to some small equipment, the biggest and the heaviest equipment 'the Support Structure' to accommodate the complete NSSS (Nuclear Steam Supply System) was being fabricated at Walchandnagar. This 'Support Structure' covers the whole reactor compartment at the level of upper deck in other compartments.

ATV established a design office in Hyderabad, DMDE (Defense Machinery Design Establishment) for the design and procurement of the secondary system for the propulsion plant, Commodore I C Rao (later rose to the rank of Vice Admiral and became the COM, Chief of Material in Navy) became the first Director. The main equipment for the secondary system is the propulsion turbine, condenser, turbo generators, main shaft and the propeller; most of the equipment was fabricated by BHEL in Hyderabad and in other locations. The submarine hull was fabricated by L&T at the Hazira works, the follow up was done by the naval engineers from ATV and the designers DND (Directorate of Naval Design), the Submarine Design Group of the Navy, headed then by R S Chaudhry and later by R B Verma,

In any organization, there are personal problems of the employees which need to be tackled. One such problem arose in RPD BARC; one electrical engineer Anupam Sharma who had joined in 1984 after completing training course from the BARC training School came to me and requested for a transfer to Kalpakkam as the civil works had already commenced and the project has taken off. Normally all the engineers, in particular those from North, like Sharma, wish to be posted only in Mumbai; thus the request was rather surprising. He then explained to me that his batch mate and girlfriend Jayanti has been posted in IGCAR and they wish to get married and settle down together. Director Engineering Services of IGCAR was very happy to have one additional engineer to look after the PRP works. He became the first person from the project to be posted in Kalappakam. When he traveled, there were very heavy rains in Tamil Nadu, he could not get a transport up to Kalpakkam due to flooding. He put his luggage on his head and waded through knee deep water from

Mahabalipuram to his destination, a distance of about 15 kilometers. Any way for love, persons have gone through much bigger hurdles.

The design work was in full swing in RPD, Reactor Control Division and Reactor Physics at BARC and the progress on the civil works at PRP was also moving ahead with the same speed; so was the case with the design of secondary system components at DMDE Hyderabad. In addition to the work being carried on the high pressure components of the reactor system, the first major order placed was for the manufacture of the 'Core Barrel' on G R Engineering Works at Bangalore. This is a cylindrical component with grid structures housed inside the reactor pressure vessel to accommodate all the fuel assemblies. This was the first major nuclear component delivered to the site.

While the civil works at the site were getting completed, the location of the residential complex about 8 kilometers away from the site and also away from the existing residential complex of DAE was selected as the land could be acquired from the state Government. V Krishnamoorthy did all the running around and interaction with the Tamil Nadu Government officials to acquire the land at Amaipakkam and started constructing the accommodation for all the DAE employees and the defense personnel working for the project. There was a separate wing for the serving naval officers and men as some additional facilities like the mess and the place and equipment for physical training was added.

It was realized that the time had come to move a team of engineers from BARC to the site, to set up the site design office to design and establish fabrication facilities required for erecting some test faculties and also to receive and erect the equipment which would be reaching the site. Thus in March 1989, S Basu was designated as the Engineer-in-charge of the site and he moved along with some young Turks including S Bhattacharya, K V Ravi, A C Bagchi, K V Krishnamurthy, V G Joshi and K Satish to the site and established the site office. Anupam Sharma and M M Rajput had already gone earlier and joined the group. Except Basu who had joined in 1975, all other engineers had just started their professional careers after studies only in 1984 and 1985. There was not enough housing accommodation in the DAE residential complexes, Pudupatnam and Sadras to accommodate this team of engineers temporarily for 2/3 years till the PRP residential complex at Amaipakkam gets completed. Thus all these engineers hired the apartments in Adyar and started commuting to the site in a hired minibus, over one hour journey in both directions. Commodore Krishnaswamy and his office was a great help for the logistic support. After about 2 years, S Basu, being the only family man and the senior most, was allotted a flat in the DAE residential complex. This continued till early 1993, when some houses in the new complex got completed and all moved. The lead was taken by Captain Nadaph (he later became Vice Admiral and the Programme Director of ATV) who was posted to the site from the ATV with a designation of PM, PTC (Project Manager, Prototype Training Centre) to liaise the activities required at site for the secondary system of the propulsion plant. In the Navy, the PRP had a different nomenclature; it was always referred to as the PTC, as it would eventually be used for training the crews before being sent to the sea on submarine. There is a lot of cultural difference between uniform-donning personnel and civilian scientists; we imbibed the best of both worlds and worked together; the motivating force being 'thrill, challenge and the prestige of the project'; this was the unity in diversity. Some high dignitaries stared visiting the site and stayed in the 'Mess' and after the technical programme, participated in some cultural and sports activities in the residential complex. Once, then CNS Admiral V S Shekhawat visited along with DG ATV Admiral Bhushan and played a game of tennis with Anand and Nadaph; all the families surrounded the court and cheered the teams.

It was recognized that all the machinery, being first of its kind, must be fully tested before it is installed inside the hull as it would be a herculean task to repair in situ or to remove it outside for any repair. MTC, the Machinery Test Center, was set up by ATV/Navy, in Vishakhapatnam to test all equipment from the propulsion turbine onwards including the propeller and the dynamometer. A boiler of requisite capacity and the pressure & temperature was installed for steam supply, replacing the nuclear reactor. The main rotating equipment of the primary system which needs to be tested is the PHT pump, (primary heat transport pumps) which circulates the water inside the reactor pressure vessel. A PHT pump test loop was also erected in the MTC. The PHT pumps have a unique history; these were first ordered on a company and when all the components were fabricated at the manufacturer's works and were ready to be assembled, due to some political reasons, company refused to sell these pumps. A number of meetings were held but the company could not be persuaded to assemble and supply these pumps; we cashed the bank guarantee and started looking for an alternative. We were successful in finding an alternate manufacturer for these pumps but the design was different with the result that we had to modify the 'core barrel' which was already delivered to the site and also had to do rework on the steam generators being manufactured at BHEL Tiruchy.

Though, the PHT pump test facility was being set up at Vizag, as the project progressed, we soon realized that there were a number of additional systems and sub systems for the reactor which needed functional testing. Thus a new high rise building, number 18, was quickly designed and constructed at the sight; the facility for testing the control rod drives, needing a large head room was set up in this building. The design of the control rod drives has another unusual story; it needed a set of bearings to be able to withstand the hostile environment. The designer Y K Tally sent an inquiry to the vendors; when he did not get any response, he went to the hardware market and visited all the shops keeping the stock of various bearings. He found the bearings of the desired size which from the look were indicative of meeting the desired specifications; these were being used in an automobile. He bought these bearings immediately and successfully tested them; his happiness showed it all, on his face whenever he described the incident. The C&I systems (Control and Instrumentation) were designed in Reactor Control Division under the leadership of G P Srivastav & R K Patil and some of these were also fabricated in house under the leadership of G P Srivastavs who had also built the simulator in house. All the large control panels for the reactor were ordered on ECIL (Electronic Corporation of India Ltd.) a public sector undertaking in DAE. In our opinion, PRP work in ECIL was not getting the required priority as ECIL mainly manufactures a large number of systems and components for the Defense. When during the same time, G P Srivastava was appointed the CMD of ECIL, he told me that he has been shunted from BARC and posted to ECIL so that the PRP work gets the required priority.

Dr. A P J Kalam, known as the 'missile man' (he later became the President of India) had taken over from Dr. Arunachalam as the Director General of DRDO and the Scientific Advisor to the Defense Minister in 1993, after Dr. Arunachalam left for the US to join as a teaching faculty member in the Carnegie Mellon University. Dr. Kalam used to call DGTV Admiral Bhushan and the Project Director PRP, Anand for briefing on the program before his meeting with the Defense Minister. This program was always included in the agenda for the periodic meetings with the defense minister. After the change in the Government, once the defense minister was from UP (Uttar Pradesh) which has Hindi as the state language; medium of instruction in most of the schools and even in the universities is Hindi. Unless one has attended an expensive private English medium school, one has a very elementary knowledge of English language; thus the minster's English knowledge was very elementary. On the other hand Dr. Kalam belonged to Tamil Nadu, with Tamil as the state language and most of the persons have very elementary knowledge of Hindi. Both Admiral Bhushan and Anand were called by Dr. Kalam when this program was discussed with the minister; they not only had to prepare the bilingual power point presentations in, Hindi and English but also had to become interpreters between the minister and Dr. Kalam; another example of 'unity in diversity' of India. Dr. Kalam used to visit the site and interacted with the engineers regularly, making us feel as if he is just one among us.

Dr. Kalam attending a presentation in the briefing room at the site, just like all others; he had a childlike curiosity

Transportation of Hull Sections to the site-- 1994

The first major equipment to reach the site was the hull sections and the bulk heads after performing the tedious journey first by sea and then by road; all the liaison work for this venture (may be adventure) was done by Captain Nadaph. These were loaded on the ship at Hazira where these were fabricated by L&T and sailed to the destination of Chennai port where these were un loaded on to a number of trailers for their journey by road to Kalapakkam a distance of about 90 kilometers. These were designated as ODC (over dimensional consignment); a reputed transporter experienced in transporting heavy and ODC by road had been selected by L & T with our concurrence. As per the transporter, the land route was surveyed and the culverts were strengthened.

The transporter had obtained standard permissions for transporting the 'Over Dimensional Consignment' from various authorities and started first convoy of three multi axle vehicles from the Port. The overall dimensions of each consignment were 15m X 10m X 12m High, weighing about 90 ton. The width and height demanded cutting of branches of trees, removal of other obstacles, disconnection of overhead power supplies, and disconnecting communication cables. On narrow roads they stared cutting branches of large number of trees; the NGO, Friends of Trees and the general public seriously objected to it and Chennai Corporation had to stop the convoy. The Transporter put all his efforts to get over the issue and restart the work but nobody gave in; thus the Transporter reported to L&T that he was stuck and could not proceed any further.

Earlier, years back, on a number of occasions, ODC were transported through the same old Mahabalipuram road during the construction of Madras Atomic Power Project and Fast Breeder Test Reactor. The citizens still remembered the hardship they had faced. The previous experience of Tamil Nadu Electricity Board (TNEB) was not acceptable; each tower cable lift and reconnection had taken one day and during that period, part of Chennai supply was off, resulting in tremendous commotion within public and loss of industrial production. If the earlier experience was replicated, the estimated time required for crossing these towers would be 18 days, provided everything goes well. This was certainly not accepted now with passage of time, in the high tech era.

Not only technical competence was required to execute the job but also the diplomatic and Public Relations skills; thus Nadaph was requested to go to the Chennai Port, assess the situation and take charge. The team prepared the detailed route map of old Mahabalipuram road, indicating areas of concern after checking the claim of the contractor about strengthening the culverts or cutting the branches of the trees by him.

The report indicated that the road width was 12 – 15 meters, all the way, with a large number of trees all along; the branches were still coming in the way of the consignment; these branches needed to be cut. Large number of overhead power and communication cables were at a height of 6/8 meters and needed to be disconnected and reconnected. At many places, 11 KV power supply cable towers crossed the roads with cables hanging at a height of about 10/12 meters. High voltage cables would be very close to the consignment or even touch and hence the power supply would need to be disconnected and huge cables would need to be lifted to allow consignments to pass under. There were 4 such convoys to be passed with a gap of two days in between for the trailers to return for next consignment. There were a large number of culverts and bridges in a stretch of 90 kilometers. Some of these were very old and some were narrow. So some need strengthening and in some cases side walls needed to be broken and reconstructed. The road in some areas was bad and in some places there were sharp turns not suitable for such long vehicles.

The operation called for tremendous co-operation of State Government, Chennai Municipal Corporation, villages, towns on the way and public at large; hence the Chief Secretary of Tamil Nadu was approached for help and co-operation from all the State Authorities. The route map and the anticipated difficulties on the way were explained to the Chief Secretary; he called for an extraordinary meeting of Chief Engineers responsible for electricity, telephones, roads and also the boss of Chennai Municipal Corporation, Conservator of Forest, DG Police and some prominent citizens of the areas concerned including the NGO Friends of Trees.

The biggest challenge was minimizing shut down of power supply. The team worked out plan which used mobile phones; these had recently appeared in the market and were distributed to local team leaders of TNEB in power distribution network and field teams engaged for lifting cables. Disconnecting power supply and communication was only done during actual crossing of the convoy and reconnection was done as soon as the convoy passed; each convoy passed all the cable crossings in one day, with disruption time at each crossing for about half an hour. Thus the disruption happened only for about half an hour in any given area on 4 days which was not abnormal for the residents. Cutting of the tree branches was coordinated with the forest Department and local Panchayat / Municipality. Finally the hull sections reached the site and were unloaded inside building number 1 with the help of the 100 ton crane.

Typical Hull Section with overhead cables.

Mazagon Docks

Mazagon Docks had been engaged in the manufacture of the conventional submarines and were invited to be yet another partner, for fitting out the secondary equipment, do the piping and weld hull sections. They set up their own facilities at the site and the manpower was housed in our residential complex. All the remaining jobs including the fitting out the reactor compartment were executed by BARC. Thus there were a number of agencies from the Government, Public Sector, Private companies and local contractors doing work at the site on 'need to know' basis; the credit for coordinating goes to Basu the engineer in charge of the site.

By the time the hull sections arrived for the land based unit, the infrastructure for doing similar jobs and much more for the first submarine was being created at Vishakhapatnam. An Admiral from the navy, who was then posted in ATV at that time, proposed that the command and control of the land based unit PRP should be handed over by the DAE to the Navy and the CO of this unit should report to the C-in-C East, Commander-in-Chief, Eastern Naval Command. His reason was that the PTC, the prototype training center would eventually be used for training the crews before being sent to the sea on the submarine; thus it should be a naval establishment though located in the DAE Complex. (Similar discussions had taken place in the beginning of the project but for better or worse, the land based unit always remained under BARC and was progressing well). Admiral's reasoning was, that there was a need to train the naval officers in supervising the fitting out before they would be called upon to do so at Vishakhapatnam, without realizing that the officers in Navy move to another job every 3/4 years as a policy as compared to the engineers in DRDO and DAE who continue in the same/similar activity for decades. In the government there are two policies for promotion; one is referred as 'vacancy based' which is the case in Navy/Defense and most of the other government departments; the other scheme is referred as 'merit based' which is the case in the scientific departments like DAE, Space and DRDO; this is the reason for the scientists and engineers in DRDO and DAE to continue in the same job for years and become specialists rather than general engineers. The matter became quite serious having a number of implications some real and others imaginary between the DAE and the Navy; then CNS Admiral Vishnu Bhagwat called a meeting on the subject and in addition to his team of senior naval officers; he invited Admiral Bhushan, the DG, ATV and Anand, the Programme/Project Director, BARC to join the meeting to express their views. Later, Dr. Kalam and Dr. Chidambaram also called Admiral Bhushan and Anand and like the CNS, they only listened without giving any comments or interruption; they also visited the site. My team and I do not know what transpired among the Navy, DRDO and Atomic Energy chiefs but we felt a sigh of relief when the decision was taken to maintain the 'status quo' after examining all the pros and cons; in any case the reactor compartment even in the first submarine and subsequent ones would be fitted out by BARC personnel who have the experience of handling nuclear activities. A new BARC complex would be created in Vishakhapatnam to cater to the needs of propulsion reactor development program as was done for nuclear power reactor development programme in Trombay.

From the 1997 onwards, the nuclear components from BHEL Tiruchy started reaching the site. The first to reach was the Reactor Pressure Vessel RPV followed by the first SG, Steam Generator in early 1998.

Reactor Pressure Vessel Reaches The Site

Though being a designer myself, I believe that doing a design of new complicated equipment or a system is easier than fabricating and testing the same to meet the requirements. As an example one can design a high pressure vessel and come out with a thickness of 100 mm of a very high strength alloy steel cylindrical vessel of a meter length but it is not an easy job to cut, bend and weld the same and to give the final shape and test it too. Nuclear components fall in that category. A B Mukherjee describes the thrilling moments when the first SG was hydro tested at BHEL.

The First SG, Steam Generator
(A B Mukherjee)

In the latter part of the twentieth century, the Indian industry had started manufacturing components for pressurized heavy water reactor

(PHWR). This technology operates on a pressure range of 100 bars; thus the steam generator for PHWR was designed for 100bars. We however were involved in the development of pressurised water reactor (PWR) technology which needed a much higher design pressure of about 200 bars, 200 times the atmospheric pressure. The SG has helical coils inside a shell of about 700mm diameter and 2 meter long.

When the first assignment was given to BHEL, Tiruchy, the Indian industry giant in south India, for the manufacturing of steam generators for PWR, it was a learning period even for them.

After a few initial hiccups, the first steam generator was finally ready for the hydro test, somewhere around 1995 to a pressure of 250 bars at BHEL. There was a lot of excitement as well as apprehension among the engineers. Everyone wished the test to succeed the very first time. Everything was being done slowly and cautiously, each step being checked by a team of experts. Designers introduced a number of additional hold points to check the healthiness of the system at each step. Extra precaution was being taken; this was the first time such a high pressure test was being carried out on such a large and critical component. In spite of all the preparations, Murphy's Law prevailed: what can go wrong, will go wrong.

We had checked the Steam generator at 50 bars pressure; everything was fine again at 100 bars and 150 bars the hydro test was running smoothly. Once we crossed this stage, everyone was jubilant, as this exceeded the earlier limits for the PHWR components. But the moment we crossed 176 bars there was a deafening 'THUD' sound, shaking us all; a dreaded event for which we were not at all prepared; everyone panicked and started checking everything. Was there any failure in connecting pipes or collectors or temporary joints? Nothing could be located. The worst fear was coming true: the failure was somewhere inside the steam generator making the situation more complicated.

After some discussion, it was decided to start opening the top flange to make way for inside inspection. To our delight, we saw that it was the temporary gasket that was faulty: it was in a shattered condition. The fact that there was nothing wrong in the steam generator itself gave us all a big sigh of relief. The spiral gasket's fibers had been shattered and had spread all over. Even the steel reinforcement used for strengthening the gasket had broken down under such high pressure. In fact, from a design engineer's point of view, it was a unique opportunity to study the effects of extreme pressure on the gasket, and thus we proposed to take a photograph of the shattered gasket.

But BHEL strongly reacted and was of the opinion that we need not celebrate a failure. It showed the improper selection of gasket and could not be projected outside. In search of a solution, we contacted a reputed gasket producer, located in Chennai. The company was explained in detail about the failure. Fortunately, even though the company had no previous experience of designing gaskets for such high pressure, they rose to the occasion and the modified gasket was supplied the very next day. This time, before we started, the test setup was re jigged to ensure that every temporary joint or connection was tightened properly to avoid any failure. When the test started, steps were repeated in the same order, but even more carefully than the first time. It did work this time. When the pressure finally crossed 175 bars, we all broke into loud raucous cheers. The happiness and relief was unmatched: the atmosphere was incredible. And the hydro test continued to work smoothly, eventually meeting the requirement of 250 bars. Thus the first SG reached the site in early 1998 after a formal flag off by the Chairman Dr. R Chidambaram.

By the time India tested her nuclear weapons in May 1998, the project had made considerable progress; the installation of the secondary system consisting of propulsion turbine, condenser, turbo generators, shaft and the propeller was nearing completion. The equipment for the reactor compartment had started arriving at the site. The Prime Minister Atal Bihari Vajpayee after declaring 'No First Use' visited the site developing the 'Second Strike Capability'. Briefing the PM Vajpayee remains the most cherished moment of my life as I have a great admiration for him since his young days and also mine. I still remember the day when I glued to the radio when he addressed the UN as our foreign minister. In my personal opinion, he is the best PM we ever got, having such humane qualities. I was very lucky to share the table for lunch with him in the afternoon at Kalpakkam, along with some other VIP's; I was feeling very conscientious of the fact and little surprised at myself. When the waiter started serving the dishes, the PM very politely told the waiter to bring 'katori for the daal' exactly like many of us do; this put me at ease and I enjoyed my lunch looking at and admiring Shri Vajpayee, all the time sitting diagonally across.

I feel the visit of the PM gave a big boost to the programme and the work for the first submarine speeded up. Along with Dr. Kakodkar, then Director BARC, I went to Vishakhapatnam to select the site for the refueling facility for the first nuclear submarine and the future ones. The decision was taken to set up the high pressure and high temperature 'critical facility' in BARC campus in Trombay to conduct Reactor Physics experiments and calibrate the cores for the propulsion reactors; the planning of the new BARC center in Vishakhapatnam was taken up after a detailed survey of the coast north and south of the existing Naval Establishment.

Author with Prime Minister AB Vajpayee

Light at the end of the Tunnel

After the hull sections arrived and the work on fitting out the equipment started, our confidence level built up and we started seeing the rays of light, though far off, at the end of the tunnel. This was the time to take stock of the fissile material required for the fuel assemblies for which the design had been completed and the development work was in progress at the AFD, Atomic Fuels Division BARC, with natural uranium. B Bhattacharjee who had taken over from Mr. R K Garg, was the Director of the 'enrichment facility' at Mysore. The total requirements and the time frame for the initial cores and refueling during the life span for PRP and the submarines was projected to Bhattachaarjee. The technology for enrichment and the facility had been a big challenge and was developed many years back keeping the future requirements in mind and thus needed substantial augmentation of the annual throughput. There was a need for immediate procurement of the sophisticated equipment and thus financing, which had to come from the defense R&D like all other funding for the ATV Programme. Dr. Kalam wished to personally asses the ground reality; I went along with him to the plant in Mysore; we rode a chopper from the Bangalore airport. After Dr.

Kalam got satisfied, no time was lost by Bhattacharjee to order the additional centrifuges and other equipment. However for the initial core for PRP, we had to wait and accept the prevailing rate of production.

Mr. R K Garg in his own words describes below the challenges faced in developing the technology of uranium enrichment.

Uranium Enrichment

(R K Garg)

When the plans of the department of Atomic Energy were being prepared for the decade 1970-1980, it was proposed that preliminary studies may be initiated on uranium enrichment as well; by this time the Plutonium plant had been working in BARC since 1965. To start with, a project team was constituted in 1970 with three fresh engineers from the training school of BARC. Over a period of 6-7 years the team size increased to about twenty engineers and scientists in various disciplines, all fresh from the training school.

Based on R&D efforts in BARC, processes for extraction of uranium from ores, conversion of the uranium concentrate to metal or oxide of nuclear purity, fabrication into fuel elements for use in research as well as power reactors and for extraction of plutonium from spent fuel, had been developed and production plants had been setup or were in an advanced stage of construction. Processes for heavy water production were also under development. Thus, indigenous technologies had been developed for all the steps of the fuel cycle except uranium enrichment. It was thought that uranium enrichment was a difficult technology and highly capital and energy intensive. At that time enrichment plants were operating only in the USA, Russia and France. These plants were based on the 'Gaseous Diffusion' process, using uranium hexafluoride as the gas for isotopic separation. Around this time some reports appeared about another process viz. 'Gas Centrifuge' being developed jointly by UK, Netherlands and Germany. In Germany 'Nozzle Separation' process was also being pursued. Difficulties in the separation of isotopes of an element arise due to the very small difference in the physical and chemical properties of the isotopes or their compounds more so for heavy elements like uranium. A very large number of stages are required for meaningful separation. Studies on separation of some of the isotopes of light elements like Hydrogen (for Heavy Water production), Boron and Nitrogen were already in progress in BARC. These processes were based either on difference in the physical properties or the chemical exchange behavior. A small plant for separating deuterium and hydrogen by electrolysis of water had been operating at Nangal as an adjunct to the fertilizer plant.

It was decided to simultaneously carryout literature (whatever little was available) survey and feasibility studies of all the processes reported to be in use for production and those under development in other countries of the world. The processes in these categories were the 'Gaseous Diffusion', Gas Centrifuge, Nozzle Separation and Chemical Exchange. Separate groups were constituted for study of these processes. For the first three processes the feed material has to be in the gaseous state viz. uranium hexafluoride. So a separate group was given the task of developing the technology for its production. It involved production of fluorine gas as well, since it was not available in the country. The technology for production of uranium tetra fluoride was already established.

On the analogy of deuterium enrichment (for heavy water production) by chemical exchange, experimental work on the chemical exchange process for uranium enrichment could be started immediately. However, after trials with a few systems over a period of about 2 years no meaningful results could be obtained. The separation factors are perhaps quite low, requiring a very large member of stages to get enrichment which could be detected by the mass spectrometer available in BARC at that time. This approach was, therefore, not pursued further.

The other three processes based on the small difference in the physical properties of the molecules of the isotopes ofU235 and U238 require, first, the development of the separating element (diffusion membrane, nozzles with very small opening or very high speed rotating cylinder). Each one requires different materials and fabrication technology. Other equipment and components required for these processes include compressors / vacuum pumps, valves, instrumentation, piping, heat exchangers etc. All items have to be compatible with uranium hexafluoride (UF6) which is a corrosive gas. Moreover, in view of the properties of UF6, the whole system has to work under sub-atmospheric pressure, and is required to have high level of leak tightness. Studies on the three processes continued over a period of 7-8 years to evaluate and assess the technical feasibility of making the separating elements based on laboratory scale work, survey of availability of required materials and equipment in the country or abroad freely, special fabrication facilities, if any, to be established etc. The costs and the time frame for the first prototype and the possibility of scale up subsequently were also taken into consideration. Based on the outcome of these studies, the 'Gas Centrifuge' process emerged as the choice for further development and all efforts were directed towards it.

After preliminary studies, the following plan of action was prepared:
i) Development of the rotor assembly (centrifuge) consisting of

(a) A tube of suitable length and diameter capable of rotating at a peripheral speed of at least 300 m/sec- which means made of material with high specific strength, and having other suitable mechanical properties, resistance to corrosive action of UF6, and thin enough to reduce the power required for its rotation.

(b) High speed motor for long continuous run, and a stable high frequency power supply source for the motor.

(c) A bearing and suspension system to support the rotor and to ensure dynamic stability at high speeds.

(d) A casing to protect against crash of the rotor and to maintain high vacuum to eliminate drag loss.
(e) Provision of facilities for electron beam welding, heat treatment, passivation, plating, dynamic balancing etc.

(f) Arrangements for feeding the UF6 gas and removal of the separated streams. A conceptual design of the centrifuge is given in the diagram on next page.

ii) Sourcing or providing special materials and manufacturing facilities to produce the above components, in large numbers- initially in hundreds and subsequently in thousands.

iii) Interconnection of the units in parallel as well as in series mode through a network of pipes and valves and sourcing of other equipment like vacuum pumps, condensers and instrumentation for monitoring and control.

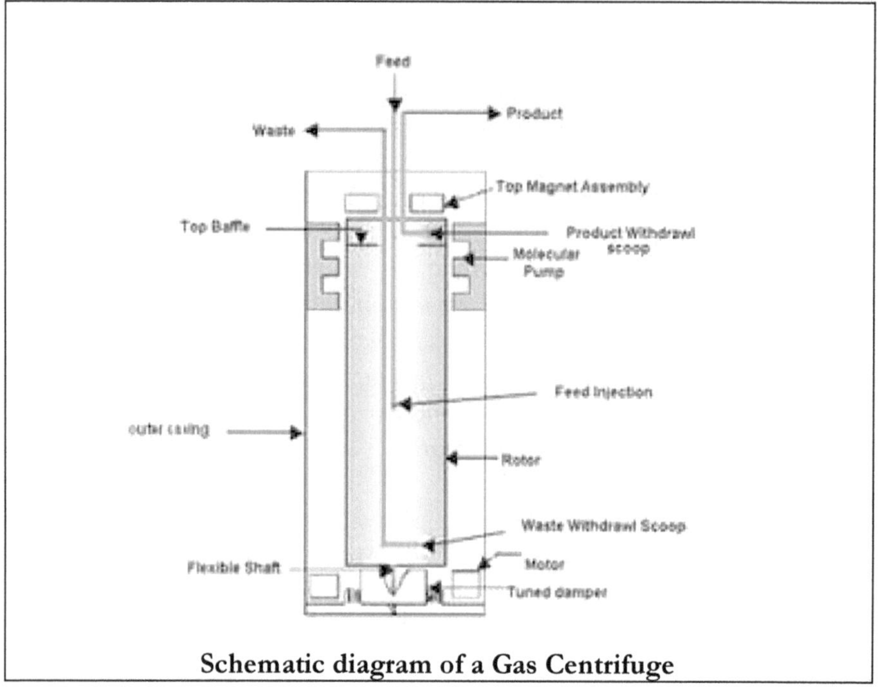

Schematic diagram of a Gas Centrifuge

From the survey carried out, it was apparent that none of the components of the rotor assembly would be available off the shelf and everything would have to be developed ab initio. Each member of the team was assigned the task of developing a specific component or system. He was given the freedom to select his line of action and to obtain guidance or help from any source he considered necessary, within BARC or outside. I acknowledge the full support of various groups in BARC and my seniors in the execution of this project.

The first important decision to be taken was on the material to be used for the rotor tube, taking into consideration the possibility of its availability indigenously or from outside sources as unrestricted supply and facilities for fabrication into tubes of the required dimensions. The candidate materials having high specific strength (Tensile strength/density) were identified as certain aluminum alloys, titanium alloys, high strength steels, maraging steel and fiber-resin composites. It was learnt that Midhani was in the process of developing certain grades of maraging steel for application in defense and the space research programmes. Maraging steel was, therefore, selected as the material of the tube. To get the tube of the required dimension, the product from Midhani had to go through two fabrication steps viz extrusion to make the blank tube of a higher thickness followed by flow forming to the required dimensions and wall thickness. It was found that at that time only one flow forming machine was available in the country, with a Govt. organization and a private party was intending to import a similar machine for making cans. The machine imported by the Govt. organization had not been installed. The private party was, therefore, encouraged to import the machine as early as possible. Thus, three organizations were identified and contact established to ensure procurement of the tube-Midhani for the maraging steel material, Nuclear Fuel Complex for extrusion of the blank and the private party for flow forming. Facilities for other operations like welding the top and bottom discs to the tube by electron beam welding, dynamic balancing, heat treatment, passivation etc. were set up in-house. The EB welding machine was developed by a separate group in BARC.

The motor to drive the rotor at a constant speed of 40000-45000 rpm was the next major component which had to be developed. The power rating was arrived at through design calculations and experimental studies. Motors with wound rotors were ruled out due to mechanical reasons. The choice narrowed down to hysteresis type solid rotor motors. Further, to eliminate any radial loading due to radial magnetic flux, axial flux motor was envisaged. Thus, a unique motor design was developed with the stator in the shape of a pancake and rotor disc directly attached to the bottom end of the rotor disc. The material of the motor rotor had to be optimized taking into consideration mechanical strength and hysteresis properties. The motor stator had also to be potted to guard against corrosion. A high frequency drive was also needed to convert the 50 Hz supply to 700-800 Hz to run the motor at 40000-45000rpm. For initial trials various sources were rigged and finally low power switching inverters were developed. One of the groups in BARC greatly helped in this endeavor.

Non-contact pickups and associated instruments for measuring the speed and vibrations were also developed, since none of these were available in the market.

The other critical component which had to be developed was the bearing to support the long rotor. The bearing had to work in vacuum, was subjected to axial as well as radial load of a few kilograms, had to run at very high speeds, continuously for years without any maintenance or lubrication. No available bearing could meet such stringent conditions. Design and fabrication of the bearing had, therefore, to be taken up in-house. After studying various systems, a hybrid design of a pivot and jewel bearing incorporating hydrodynamic action was adopted. Spiral grooves were made on one of the bearing surfaces which would generate a fluid film between the two surfaces and ensure a long life. A stable hydraulic fluid with very low vapor pressure had to be selected and procured. The problem of machining of the pivot and the jewel to the required sub-micron tolerances and engraving the spiral grooves had also to be tackled. Some enterprising entrepreneurs were roped in to undertake this job. To reduce the load on the bearing and for radial stability, a magnetic suspension system was incorporated.

By 1983 the above efforts enabled us to make the first prototype unit which could run at speeds close to the target and on which some separation trials could be conducted using some gas mixtures. This was followed by isotopic separation using the process gas viz. UF6 which by that time could be made on a small scale. A few more units were then assembled and connected in series to partly simulate the conditions of a cascade. A semi pilot scale facility was, thus, available for component and system testing and for conducting separation trials. Concurrently a decision was taken to set up a demonstration plant with a few hundred units. A new site was selected and civil construction work started in 1984. Along with experimental work in the semi pilot plant and construction work of the demonstration plant buildings, activities connected with sourcing of materials, fabrication and machining , procurement of bought out items had to be intensified.

The demonstration plant was planned to have a cascade consisting of a few hundred rotors to start with, facilities for testing of components and assembly of rotor components as well as for UF6 production. By 1989 all the construction activities and installation work were completed and the plant was commissioned with UF6 process gas and enriched uranium was produced in the country for the first time in kg quantity. A venture which appeared to be fraught with many uncertainties became a success story and a matter of pride for all of us who were participants in it. It was possible due to the full commitment and dedication of the team, their untiring efforts and innovative approach and co-operation of many groups in BARC.

When it was decided to enhance the production capacity of the enrichment plant, it was decided to set up the plant for converting enriched UF 6 (Uranium Hexafluoride) into the metal in BARC Trombay; the plant was designed and set up by Mrs. S B Roy under the overall guidance of Dr. C K Gupta, Director materials Group. In parallel, the development work for manufacturing the fuel assemblies was in full swing at AFD, Atomic Fuels Division, using natural uranium; AFD was then headed by a very dynamic engineer R P Singh. The designer K N Vyas started spending half his time in AFD. The challenges faced are best described by them.

Challenges in conversion of U Hexafluoride to U Metal Powder (Mrs. S B Roy)

Uranium Metal Powder (UMP) being the first item for starting the fuel production, first step was to develop a process for making this powder. Because the uranium metal is used in our Research Reactors, the initial trials were done to produce the powder from Uranium Metal Ingot (UM INGOT). However, it was soon realised that to get chemical purity and the required shape factor and particle size distribution it was not possible to do the conversion of INGOT to powder by chemical or mechanical means. Thus all effort was put by the Uranium Extraction Division (UED), first to develop a feasible process for making the Metal Powder of required quality and then to establish a facility inside BARC for continuous operation and production of required quantity of Metal Powder.

The chemical reduction process was chosen for the production of Uranium Metal Powder (UMP) from Uranium Hexafluoride (UHF) through wet route. The fundamental principle of reduction in this process is that after reduction reaction, all the reduced species join into forming a metallic phase and the oxidized species form the slag phase. If the heat effect of any given reaction is sufficient to melt the slag and the metal phases, the metal is obtained in a massive, consolidated form. If, on the other hand, the heat generated in a given reaction is insufficient leading to the melting of the slag phase, the metal is obtained in a powder form that remains intermixed with the slag phase. Then the metal is recovered by separating it from slag either by mechanical or chemical means.

The calciothermic reduction of uranium dioxide was chosen to produce the powder.

By repeated parametric simulation, safe reduction was carried out and the powder was forwarded to AFD (Atomic Fuels Division for conversion to inter metalloid. By changing the reactant ratio and reduction reactant temperature profile, there was a significant quality improvement in the product; however, the Oxygen content reactants played the major role and dictated the use of inert gas glove box, for handling the charge for reduction reaction and thereafter for product segregation. For product segregation, no physical segregation method was useful; thus the chemical separation method was developed. Considering the process safety and product recovery, the challenges in chemical separation process were multi-dimensional with contradictory requirements. On one side the challenge was obtaining maximum purity; on the other side was maximum recovery. Based on these requirements safe solvent like water, some inorganic acids and organic acids were tried. For higher purification and for significantly high recovery, a very low temperature acetic acid leaching was found to be the best. However the residual organic acid from the product was removed by giving another wash with Inert Volatile Solvent. The highest safety concern was observed during this stage as inter particular collision of these particles in solvent are a fire hazard; thus designing the process equipment for leaching and for the wash was an engineering challenge.

After establishing the detailed engineering of the process of Uranium Dioxide reduction by calcium, enriched Uranium Oxide (UO2) had to be obtained. The valuable enriched Uranium was in UHF form and the dry method of conversion from UHF to UO2 was under development stage; thus it became another urgent requirement for UED to engineer the wet route production of UO2 through Ammonium diuranate precipitation which was an established process for getting the purest uranium oxide. Two major developments were done in UED within a short time of 12 to 15 months. One was the process development for reduction of UO2 and the other one was engineering development of wet route production of Uranium Dioxide.

After the development, a production plant was needed. A suitable area (an existing slag storage shed) inside BARC was located to construct the plant. As the shed was meant for temporary storage of slag material, no utility services were provided in the shade and no safety features were incorporated during initial construction; process safety and nuclear material accounting were paramount. Considering these additional requirements, the dry facilities were housed inside inert atmosphere glove box and wet facilities were housed in controlled ventilated enclosure.

After completion of detailed engineering of the plant Mr. Anand extended his financial powers for early fabrication of this plant utilizing exotic materials of constructions and utilizing secret design documentation mode. One former BARC engineer Mr. Champakvas of Symec engineers extended his fullest cooperation and executed the job; the plant was first of its kind inside BARC. Dr. C K Gupta, Director, Materials Group was given the responsibility of safety review of the design and construction of the facility; criticality monitors in the wet processing area were installed. Immediately, after safety clearance, plant was commissioned without any formal ceremony and it continued operating and producing the material which was as per certified grade with the designed capacity and was handed over to AFD for further work.

Fuel Development—Challenges faced
(K N Vyas and R P Singh)

Fuel design, development and fabrication for the propulsion reactor posed many challenges and offered golden opportunities to Indian scientists and engineers to master fabrication of entirely new type of fuel & fuel assembly and led to development of many new technologies.

Conventional power reactor fuel uses ceramic pellets, Uranium oxide, nitride or carbide encased in Zr-alloy cladding. The technology was developed in BARC and the production of fuel for the power reactors is regularly done at NFC (Nuclear Fuel Complex). However, after a lot of literature survey, it was soon realized that the propulsion plant needs a different concept. There have been instances when the fuel used is a metal-zirconium alloy or uranium dispersed in some metal matrix (dispersion fuel ; even plate-type fuel, composed of enriched uranium sandwiched between metal cladding has been used. This fuel type results in a very rugged fuel, necessary for the military operation. Thus we zeroed down to dispersion fuel which also results in low fission product release in case of failed fuel. In the close arrangement of equipment for a nuclear submarine, lower spread of contamination is considered essential for lower radiation fields.

Dispersion type fuel was developed where enriched uranium in intermetallic form is dispersed in a conducting matrix. Production of uranium intermetallide of right composition, stoichiometry, phase and size range was a challenge. As described by Dr. (Mrs) S B Roy, a plant was set up by UED for production of uranium metal powder used in synthesis of uranium intermetallide. The design of fuel requires a very close tolerance on fissile content; this was a big challenge to fulfill as granules from large number of batches having variation in heavy metal composition need mixing. In addition, the requirement on uniformity of fissile atom distribution in the active length is also very stringent. These two aspects required large number of experiments.

For ensuring high reliability during operation, it is necessary to have an elaborate inspection required for confirming uniform distribution of dispersant and dilutant; ensuring soundness of bonding and measurement of dimensional accuracy. As the fuel type was new and first of its kind, being developed in the country, and it needed development of new inspection methodology and manufacture of new measuring equipment. Each and every fuel and core component was examined by experts to ensure the stringent quality control requirements.

The core of the marine propulsion plant is required to have a long life. This requires relatively large fuel inventory. As is well known, large fuel inventory results in high core reactivity and this needs to be suppressed using burnable poison. Different burnable poison materials were investigated and finally used in core.

All the new technological requirements needed a new facility for manufacture. The facility was constructed in a very short period in BARC to meet the overall project schedule.

The CNS Admiral Vishnu Bhagwat examining the first prototype fuel assembly at AFD, BARC.

The Reactor compartment has bulk heads on both sides and these have lead and concrete shielding and during the reactor operation, no entry to this compartment is permitted. In spite of the shielded bulk heads, during reactor operation, the radiation field is high due to water activity in the reactor compartment; however because of the bulk heads, the adjacent compartments on both sides can be entered during operation, to maintain the equipment. The main equipment in the Reactor Compartment is the 'Support Structure' covering complete length and width, to house the complete NSSS consisting of the RPV (Reactor Pressure Vessel) and the Steam Generators, Pressuriser, PHT pumps and some other equipment.

The sequence of assembly of the NSSS and the 'support structure' was discussed and debated in detail in BARC among a number of engineers. The first alternate thought of was similar to the one used in power reactors; to build the support structure in the hull at the required level, install the equipment and do the piping and wiring. It was soon realized that the fitting out inside would not be possible because of space constraint. The second alternate was to fabricate, do the final machining of the support structure, install inside the hull either by inserting from the side or accept a longitudinal weld and then install the NSSS. If the longitudinal weld was acceptable the installation of the NSS will be simple. The longitudinal weld was ruled out by the designers. It was finally decided to get the fabricated and machined 'support structure' from the shop, install the NSSS and then role the complete assembly from one end; later do the piping and wiring inside.

Thus the 'support structure' was being fabricated at Walchandnagar and was getting delayed much beyond our expectations; as I wrote earlier that it is easier to design a complicated equipment, system or structure than to fabricate the same with desired accuracy. A B Mukherjee was requested to camp at Wachandnagar to supervise the final machining, get it packed and dispatched. He described in his own words the challenge faced at the final machining.

Challenges in design and fabrication of ST
(A B Mukherjee)

The 'Support Structure' is a complicated structure made of SS cladded high strength steel with a number of penetrations and attachments, some of these are made of SS. The overall dimensions are about 6mx6m and a height of about 4 m weighing about 100 tons. It needs final machining after assembly with very tight tolerance, before packing and transporting to the site. This houses the NSSS. Walchandnagar Industries were selected to do this job after evaluating their fabricating, machining and handling facilities; the engineering back up was provided by us from BARC. We even posted our engineers there not only for inspection but also for providing guidance in manufacturing practically on the daily basis. In addition, the senior persons from our design team used to visit the site periodically to discuss the midcourse corrections on engineering and manufacturing problems as this was the first time that such equipment was designed in the country.

It took much more time to manufacture than what we had estimated; there were a number of problems faced during manufacture, I will describe some...

Final machining was an excited moment due to its large dimension and the requirement of tight geometrical tolerance. The final machining was planned in CNC lathe which was kept in controlled temperature to prevent temperature effect on the job, which can be otherwise significant considering the dimension of the equipment.

However, when the expensive machine with a capacity to reach long distance reached the near extreme position of the job, it started vibrating with a deafening screeching sound from the cutting tool because of the excess feed. Mr. Sade, the engineer in charge knew that job had been inordinately delayed and a number of revised delivery dates had been promised and the latest committed schedule must be adhered to. After discussing with the operator, he concluded that there was some scope to reduce the feed. With a new resolve, Mr. Sade gave the instructions to reduce the feed to 0.5mm/min. The machine still vibrated, and the resultant noise was ear-splitting; bad enough to show Mr. Sade it was not satisfactory; the minimum possible feed of100 micro meters/min was fixed; the machine did perform well; it reached its target covering the full width of the structure without any untoward noise or vibrations.

Mr. Sade knew that with this feed, the target completion date would exceed the deadline, and delivery would be further delayed. Realizing the seriousness of the situation, he approached the director and told him about his predicament. "We have to work within the capability of the machine, which can at the most reach 5m and thus, calls for two settings of the job." The Director of the company finally asked Mr. Sade to interact with the designers, to take their views about the possibility of two step machining. On detailed assessment the designer agreed for the two step machining with a rider that the mating parts also needed to be re-machined so that designed net deviation was maintained.

Welding of cavities was another problem. Unfortunately during the fabrication we lost one welder while working on a cavity. It was a deep cavity of about a meter diameter and about 3m deep. As per the manufacturing sequence, substantial work was planned after the cavity was fitted in position with the aim to get better dimensional tolerances. It was tungsten arc welding (TIG), which was preferred, as it gives clean weld without much of repair which was a desirable requirement considering the geometrical constrains of repairs. However no one realized the lethal effect of argon gas, which is otherwise known to be nontoxic. It is heavier than air and thus while welding it started settling deep into the cavity leading to depletion of oxygen. Surprisingly the deficiency of oxygen was not felt by the welder working there and he collapsed after some time without giving any signal to any one for rescue operation. We leant it a hard way the manufacturing of ST.

This platform of 'Support Structure' after loading the NSSS equipment, weighing about 600 tons (Referred as the Reactor Assembly) was supposed to be rolled in from one side into the hull and secured with its support structure integrated with the hull section. The welding of the pipes of the Steam Generators with those of RPV was supposed to be done with great accuracy, after installation of the equipment in the cavities and before rolling in the Reactor Assembly as explained above. Due to the delay in fabrication of the 'Support Structure' at Walchandnagar, the sequence needed a change, to make up some of the lost time. Sujay Bhattacharya and Pradip Mukherjee describe below, in their own words the challenges faced in executing this job at the site.

The Challenge 'Assembling the Reactor Compartment'

(S Bhattacharya and Pradip Mukherjee)

The Reactor Compartment was the last to be completed because the 'Support Structure', being the main equipment to house the NSSS, was the last to reach the site from Walchandnagar, and was in fact, shielding all the deficiencies of supply and material planning of our project. It was taking too long and the site team was eagerly waiting and honing their skills to deliver the desired goal. It covered all aspects starting from the special shielding material, testing, concrete block making, setting parameters for curing, strength, and the neutron parametric study. A lead melting and casting facility was set up.

Before the 'Support Structure' reached the site, lead shielding work on one of the bulkheads of the reactor compartment, was completed. Around 145 mm thick lead blocks were casted and fitted in 'situ' on the lower half. To carry out this lead shielding the adjacent compartment (housing the bulkhead along with hull) was made horizontal (rotated by 90 degrees) using a specially designed tilting structure. After completion of lead shielding, a 10 mm cover plate was welded on the top. It was again made vertical using the specially designed tilting structure. It was great challenge for the young Turks. After competition of the hydro test of the reactor compartment, the other bulkhead was cut to give passage for the assembly to roll inside the reactor compartment. The lead shielding was carried out in the lower half of this bulkhead in the similar fashion as in the first bulkhead.

As the 'Support Structure' was getting delayed, the site team thought of welding 4 steam generators to the RPV in advance so that 9/10 months could be saved in the project schedule. The young team at the site not only could do 'in depth' analysis but a lot of 'lateral thinking'. As it happens, in all complicated and difficult projects, the designers send the fabricated equipment to the site for installation/welding/connection to the systems, with proper instructions and the procedure to be followed, in particular, for the critical equipment. At times, the designer himself went to the site to supervise the installation and welding. Thus there was a lot of debate for a few days, between the site engineers and the designers. The designers had specified extreme accuracy for the distortion during welding, so that the stress/load would be minimized in the operating conditions, due to the differential thermal expansion in the vertical as well as horizontal plane. The final 'Go Ahead' signal was given by Mr. Anand and the work proceeded.

To address this problem, a separate 'Reactor Assembling Stand'' simulating the Support Structure with locations and accuracies required were erected in Building -1. The RPV which was shipped a year ago from BHEL, was available at the site, it was prepared and placed on the Welding Stand. The SGs started arriving one by one. The joining of RPV and SG started with SG-1.

The site team demonstrated the accuracy of welding and also subsequent placement of the assembly of RPV and 4 SGs on the 'Support Structure' after it arrives and is placed on the Rolling Structure. The SGs sit on the structure and are leveled on spring supports with desired design load on the spring supports. This was a great achievement for the site team, (S Bhattacharya, PVNL Narashima Rao, Neeraj Uttam, Ranjit kumar, DVV Prasad, Srinivasan) probably never attempted earlier in BARC. The welding distortion was estimated in each weld bead and sequenced accordingly to get the desired spring support loads on the SGs;

it is epitome of highest engineering, as weld distortions are very sensitive to heat input and sequencing. Now the imagined monster appeared "welding defect" barked at our door step; step by step radiography and load correction steps were introduced to reach the goal. We remember one adage "Life is full of thorns" but Indian wisdom from the Bhagvad Gita "Karmnyee Vadhikaraste Ma pholesu kodachana" saved us.

In the middle of July, 2001, the 'Support Structure' finally arrived from Walchandnagar by road. The story of journey can be best told by the contractor how he enticed all the authorities to pass the consignment. It reached around 5 pm at Kalpakkam. It was stinking like hell, as all the cavities had collected dirty water, from the road sides, due to heavy rains; this water splashed and filled the cavities. After thorough cleaning and the preparatory work, it was lifted and placed on the 'Assembly Rolling Stand' and the required shielding work at bottom began. Concrete was poured in the cavities as required by the design and measurement of cavity depths was done with high precision taking the help of computerized drafting. The concrete was cured for 7 days and the cover plates were welded with meticulous planning and sequencing; 'hats off' to the welders who could get into the cavities and do welding.

The rolling stand was innovatively designed, to be handled by the existing 100 ton overhead crane. The support at the hull was designed for a maximum load of 1000 tons, and the "Knee" at hull was leveled within 0.05 mm accuracy. The work related to the support was completed well in advance. After placing the Structure on the Rolling Stand, the trial rolling of the empty 'Support Structure' was done before installing and welding the equipment.

The last SG and other equipment were joined to the RPV 'in situ'. By this time the site team was well versed in all kinds of weld distortions, alignment etc. The welders did marvelous job by doing welding in very difficult positions, in all odd locations. The shutdown cooling pump designed to be connected to outlet plenum through Purification Heat exchanger, could not be placed as the cavity dimension was less than specified. Then the help came from the "in situ" machining technique. Finally the pump was installed and related piping was completed, before placing central shield weighing about 12 tons on the support structure with desired alignment.

The biggest jigsaw puzzles the designers gave was Assembly shielding. More than 60 shield blocks with various shapes needed to be placed in the vacant places. The site team first made the layout on the floor and located the shield blocks keeping the place for piping.

The next challenge came when steam flow, feed flow, blow down and sampling lines had to be welded to the SGs as the assembly gets camouflaged when it sits on the structure. Meticulously all these lines were welded to the SGs. The valve assemblies in the steam chest area consisting 3 sets of motorized and manual steam valves in port side and 2 sets of manual & motorized steam valves in starboard side were welded before carrying out the seam welding of the compartment. After completion of the hull welding the challenge was to weld the valve connector with the bulkhead penetration. Hardly 100 mm space was available at the bottom of the penetration. However our welder Srinivasan did a marvelous job by carrying out the welding keeping himself in upside down condition. I don't know if anyone has done welding on this type of constraints ever before, defying 'gravity'.

After all activities were completed, the day came when the rolling of the Assembly into the reactor compartment had to be done One day prior to the rolling of the Assembly, we noticed the expansion gap between the Assembly and the bulkhead was less than design value. The rolling of the Assembly was postponed by 15 days. The site people with zeal grinded the shielding cover plates of the bulkhead and the central deck shield blocks, to make 60mm clear gap. The Assembly was rolled inside. It was historic day and more than 100 persons watched the aggregate rolling inside. We saved almost 2 years of project time by slicing the chain activities and advancing some unrelated activities.

Subcritical experiments in the Spent Fuel Storage Pool

The reactor was ready and the fuel assemblies started reaching the site, ready to be loaded. The team of reactor physicists under the leadership of Dr. Dwivedi had been refining their calculations over the years; any time they found, in literature, an experiment performed even remotely matching with our design of the core and fuel assemblies, they would do the bench marking of their computer codes, in particular they are always worried about the temperature coefficients of reactivity, excess reactivity and shut down margin. But it is almost impossible to find an identical geometry, thus it is a practice to do critical experiments in a facility for a new core configuration because of the different fissile content and other elements over a cross section. The critical facility being built at BARC Trombay would take a few years to be ready for testing the fuel cores for the submarines; the computer codes used were based on the operation of the Tarapur boiling water reactors, whereas this reactor was a pressurized water reactor with higher enrichment. Dr Dwivedi opined that at least we should do some 'subcritical experiments' in a pool of water to build up the confidence level. Though the experiments were performed just after Dr. Dwivedi and I retired, Dr. Dwivedi describes below the challenges faced for the Reactor Physics Design in his own words.

Challenges in Reactor Physics Design
(Dr. S R Dwivedi)

If I remember correctly, it was somewhere in the beginning of eighties that we were all having a meeting in the ECIL guesthouse in Hyderabad; Dr. Dastidar , the Director Reactor Group asked "Dwivedi, can you make a reactor within a diameter of 100cm and a height of similar order to produce 100 Megawatt fission power?" Similar questions were then asked to the design engineers whether they will be able to remove so much of heat by pressurized water and able to control the reactor. We had come prepared and that enabled us to answer in affirmative; for the last many years, we as a team, had been working on precisely such a reactor, though only on paper, doing parametric studies by changing various parameters. I can safely assume this as the time when the work on building a Submarine Propulsion Reactor was initiated.

In reminiscence, I sometimes feel foolish on my part to display such confidence all by myself. However, in hindsight, I was never alone; a group of about 30 scientists in the Theoretical Reactor Physics Section, headed by Mr. B P Rastogi, were there to work on the reactor physics for the propulsion reactor; it was a very big opportunity for our group to prove our capability and the team work.

A similar situation had taken place a few years back; a working group was formed to develop the fuel management service for the Tarapur Reactors. The meeting took place in the Nuclear Fuel Complex, Hyderabad. There were members from other sister organizations, however, the reactor physics calculations could be performed only by our group or the Reactor Physicist of the Tarapur Power Station. The question too, was similar as to whether I can provide the fuel Management service; my answer was a simple "yes", provided responsibility of the supplementary thermal hydraulic calculation is owned by Mr. A.K. Anand. Reactor Physics and Thermal Hydraulics depend upon each other; at times these could even be coupled in the same computer code.

The reader may ask, why Mr. Anand? This story is equally interesting; Mr. Anand and I had been deputed, for 6 months, to the General Electric company in San Jose, to participate in Reload Fuel Design for the Tarapur Reactors and our job was to learn everything about the design and the fuel management of BWR (Boiling Water Reactor, Tarapur). Before the PNE in 1974, we had provided FMS for two cycles under the guidance of GE; now we had to be completely on our 'own'.

When these Tarapur Reactors, were installed, it was agreed that USA will provide the enriched fuel for the reactors, but will not take back the irradiated fuel. India will be responsible for keeping the irradiated fuel safe underwater.

In the agreement, USA had agreed to train our scientists in the Battle Northwest Laboratory run by the US Government for Plutonium Utilization for one year and I was deputed to Richland, Washington for one year in 1969, to learn about Plutonium Utilization. After completing this training, I was asked to spend 6 months in General Electric at San Jose, to join the group of GE scientists working going on for Reload Fuel Design of Tarapur Reactors.I was made responsible for Physics Design and Mr. Anand was then deputed to join this team for Thermal Hydraulics and mechanical design of reload fuel; this is how I met Mr. Anand for the first time, at San Jose; and that is how the dots connect, you see!

Anyway, to progress the story further, in BARC too, to utilize Plutonium, a working group was formed to find various means of using Plutonium – in supplementing the fuel for Nuclear Power Generation. Plutonium Oxide can be mixed with Uranium Oxide and this Mixed Oxide (Mox) fuel can be used either in Thermal or Fast Reactors. Fast Reactor technology was in its nascent stage and MOX was being tried in LWRs (Light Water Reactors) in other countries. We also started working on Utilization of Plutonium for power generation and the general term for this type of work was referred as Plutonium Recycling Program or (yes you guessed right!) PRP. PRP is basically the physics work involved in developing the computer codes for analysing fuel assemblies for neutron flux distribution and reactivity.

My USA experience had helped me get many experimental results of fuel assembly lattices containing different MOX Fuel with varying amount of Plutonium. Developing the computer codes and analysing those lattices was the work taken up in our own section; I became a part of this group. The contribution of this group became very important when the supply of enriched fuel for Tarapur became uncertain; 2 fuel Mox fuel assemblies were fabricated and successfully irradiated in one of the Tarapur reactors.

Though it was well known that Naval Propulsion Reactor would require enriched Uranium fuel, there was no enrichment facility yet in India and after the PNE at Pokhran, we could not expect any other country to supply enriched fuel to India for this reactor; not even for experimental purposes. There was a general feeling that if we can make an explosive device with Plutonium, then we can probably use Plutonium enriched fuel instead of enriched Uranium. Many paper designs were made for Propulsion Reactor under the Plutonium Recycling Program (PRP). I worked as the Group Leader for such physics work, and my boss, Mr. Rastogi was the member of the design team for Plutonium Recycling Project, a land based prototype propulsion plant. The code name PRP suited perfectly to keep the confidentiality of the work being done.

Our group started making physics reports for this type of reactors using Mixed Oxide fuel. Such reports gave an idea about fuel control and shielding requirements. In developing computer codes to analyse such reactors, there were two basic problems; 1) development of computer codes to handle the required geometrical details and 2) validation of these codes using some known experimental data. The nuclear data or what is called the cross-section library was also not verified for such lattices. With the available material in our command, many physics design reports were produced and discussed among ourselves. All published literature and unclassified reports were gone through to obtain experimental data to validate our code system. New computer codes were developed in the group to cater to the required geometry and to generate input data for engineers for core design.

Fuel design engineers Mr. S Basu (present Chairman Atomic Energy and Secretary, BARC) and Mr. KN Vyas (present Director, BARC) were always interacting with us for the required inputs. In other countries, the fuel design is always done by nuclear engineers having good background of both reactor physics and engineering. In India, physics is cut off from the engineering discipline, hence for a proper fuel design for PRP, both engineers and reactor physicist had to work in close collaboration. At this stage, I withdrew partially from the group.

To satisfy my academic interest, I got myself registered for Ph. D. in Bombay University and started my favourite work on Monte Carlo Studies. After, I completed my Ph.D. I was taken back again to continue on the PRP work; by this time, Mr. Anand was appointed as the Project Manager and a real BARC project was being built in Kalpakkam.

Why it is so difficult and challenging to design a small compact light water reactor, when we are able to design different types of large reactors for power generation.

Some of the problems can be broadly categorized as:

1) *Large Power density but smaller neutron flux density, to avoid xenon over rides problems.*
2) *No fuel management, single core should last as much as possible.*
3) *During the lifetime of the core, power distribution should not change drastically so that control rod movement could be avoided.*
4) *It should work like an electric heater – you switch on for generating power and a sliding switch for linearly increasing power.*
5) *During compensation for xenon transients, the power variation should be minimum and within limits.*
6) *90% of the time, the reactor will be operating at 25% of its capacity*

Actually engineering problems were more severe, like rolling and pitching, greater stresses, salt water corrosion, etc.

To understand all other problems and getting the required experience, a land based reactor was planned at Kalpakkam.

As I mentioned earlier, this project was designated as PRP and Mr. Anand was designated as Project Manager. Mr. Basu was responsible for mechanical design of the fuel assembly and we interacted regularly to achieve the best possible design. Later on, Mr. Vyas took the responsibility when Mr. Basu shifted to Kalpakkam site as the engineer-in-charge. Physics design of the fuel assembly is generally available in reference books for large LWRs with enriched Uranium Fuel, and this is how we progressed:

Fuel pins sheathed in Zirconium clad are arranged in a square or hexagonal pitch with very little space for coolant to flow through these pins and remove the heat generated due to fission. Some of these pins contain burnable absorber like Gadolinium or Boron to compensate the reactivity change during burn up.

These fuel assemblies are now arranged in such a way to make a cylindrical core. Requirement of a relatively smaller diameter led to a hexagonal fuel assembly. The reactivity of the core changes drastically in bringing it from cold room temperature to hot full power conditions. To compensate for these reactivity changes, moveable control pins are introduced in these fuel assemblies.

We then start with a tentative core design and evaluate the neutron flux distribution/power distribution during the entire life of the core.

Many hot spots are recognized which may lead to breach of the fuel cladding and leading to contamination by radioactive fission products in the coolant – and hence in the entire system. This is not permitted. This leads to us studying many core designs. We have to calculate three-dimensional power distribution throughout the core life and satisfy ourselves that the core will suffice all the control and hot spot requirements.

In naval propulsion reactor there is usually not enough space to provide a lot of instrumentation, hence there is zero tolerance for errors. Most of the experiments are performed before loading the core. A critical facility is normally available to do the experiments on the actual core; this provides the necessary data for reactor safety and operational requirements. Thus we planned some experiments, in particular to measure the control rods worth, in the Spent Fuel Storage Pool in Kalpakkam. Immediately after my superannuation, the experiments were performed by my colleagues, before the fuel was loaded in the reactor. The happiest moment of my life came when a message was delivered to me from Mr. Basu that the reactor was made critical; he also added 'as predicted by you'.

Unfortunately, I (the author) could not see the reactor being made 'critical' in my service as I superannuated in November 2001 while the reactor became 'critical' on November 11, 2003; it happened to be my birthday, as per the office records; the message was conveyed to me on phone from Kalpakkam. The credit really goes to the young team of engineers who joined the professional career from 1984 onwards, along with the start of the project and continued till date.

INS Arihant

On 26 July 2009 India launched its first nuclear-powered submarine, INS Arihant, which means "Destroyer of the Enemies" in Hindi according to the official news release. The name Arihant has its origins in the Jain religion, and unofficial news reports stating "Destroyer of Enemies" omitting the definite article. India became the sixth country in the world to have built one. The nuclear propulsion reactor on board INS Arihant, achieved criticality on 10 August 2013, paving the way for its operational deployment by the Navy; she had been undergoing trials at Navy's key submarine base in Vishakhapatnam. On 13 August 2013 Prime Minister Dr. Manmohan Singh stated "I am delighted to learn that the nuclear propulsion reactor on board INS Arihant, India's first indigenous nuclear powered submarine, has now achieved criticality. I extend my congratulations to all those associated with this important milestone, particularly the Department of Atomic Energy, the Indian Navy and the Defence Research and Development Organization. Today's development represents a giant stride in the progress of our indigenous technological capabilities. It was testimony to the ability of our scientists, technologists and defense personnel to work together for mastering complex technologies in the service of our nation's security."

ET Bureau | Feb 23, 2016, 09.33 AM IST reports-- India's first nuclear armed submarine is now ready for full-fledged operations, having passed several deep sea diving drills as well as weapons launch tests over the past five months and a formal induction into the naval fleet is only a political call away. Multiple officials closely associated with the project to operationalize the INS Arihant nuclear missile submarine have confirmed to ET that the indigenously-built boat is now fully-operational and over the past few months, several weapon tests have taken place in secrecy that have proven the capabilities of the vessel. Technically the submarine can now be commissioned at any time," a senior official said. Sources told ET that the commissioning date could be as early as next month if the Modi government desires. A communication facility to interact with the submarine has already been commissioned into the Navy.

I am sure every Indian will feel proud as the country in her 70th year of Independence, matched the 5 super powers which are the permanent members of the UN Security Council. The British naval officers, in 1958, were of the view that even the conventional submarines were far too sophisticated for Indian Naval personnel to operate!!

NUCLEAR PROPULSION IN BRAZIL

In the 1970s, when the nuclear propulsion programme for the submarines was envisaged, the Brazilian Navy decided to buy diesel-electric submarines. At that time, the international market for submarines was controlled by the French Daphne class, the English Oberon class, the Soviet Foxtrot class and the German IKL 209. The Navy, then chose the German class. The Navy signed the agreement with the German consortium Howaldtswerke-Deutsche Werft (HDW) for the construction of the first submarines IKL; the transfer of technology for the project was not intended. The first unit of this model was built in Germany in the presence of Brazilian engineers and technicians, and the other four units were built between 1982 and 2005 in Brazil, in the presence of German engineers with intensive participation of engineers and technicians from Brazilian naval force in the Navy Yard in Rio de Janeiro (AMRJ) and Nuclebrás Heavy Equipment SA (NUCLEP).

In July 2007, the lawyer Nelson Jobim took over the Ministry of Defence (2007-2011) and made a political commitment to reequip the Brazilian Armed Forces. In October of the same year, Jobim participated in a public hearing organized by the Committee on Foreign Relations and National Defense (CREDEN); in the plenary three House of Representatives, publicly advocated "the reequipping of the Armed Forces with a transfer of technology to Brazil, even if there is a need to import". In order to enable this option for technology acquisition, Jobim and military commanders approached the countries that would be willing to transfer strategic knowledge to the Brazilian Armed Forces. In January 2008, the Minister of Defense and Brazilian military commanders visited military installations in France and Russia and discussed technical and technological cooperation agreements. As France and Russia were the only two countries at that time that could build and sell conventional and nuclear submarines; these countries were selected for further negotiations by the Brazilian government. Russia offered Brazil the diesel-electric submarine Amur and France in addition to offering the new conventional submarines Scorpène class, also committed to sell the submarine hull with nuclear propulsion technology transfer. The main buyers of Scorpènes DCNS were Chile, in 1977, Malaysia in 2002, India in 2005 and Brazil in 2008. Of the four countries that France has negotiated the sale of submarines, only two can be considered strategic: India and Brazil, as in the negotiations with both the countries, the transfer of technology was a part of the agreement.

The history of business relationship between France and Brazil in defense, dates back to the early twentieth century, which includes the acquisition of assault vehicles, supersonic fighter jets and military helicopters. The involvement of the Brazilian shipbuilding industry in this partnership is more recent, but not less important. According to the National Defense Strategy (END), because of the limited resources it is not feasible to control the shipping routes, power projection and denial of the use of the sea to the enemies. Among these three strategic objectives, END determined that the Navy of Brazil prioritize the denial of the use of the sea. To fulfill this mission the force will have to provide an integrated advanced denial system consisting of aerospace defense systems, ocean patrol vessels, coastal and multipurpose patrol, fighter planes, helicopters, unmanned aerial vehicles, submarines with modern propulsion systems, autonomy and speed, advanced ballistic missile systems such as anti-ship missiles and a regional power projection. Thus, by prioritizing the denial, the Navy is also developing the other strategic objectives. The END adds that, to ensure the aim of denying the use of the sea, Brazil will feature underwater naval force, made up of conventional submarines and nuclear-powered submarines. Brazil will maintain and develop the ability to design and manufacture both conventional submarine propulsion and nuclear propulsion. Accelerate investments and partnerships needed to run the nuclear-powered submarine project. Will arm submarines with missile and develop skills to design them and build them. Take care to gain autonomy in cyber-technologies that guide submarines and their weapons systems, and to enable them to operate in a network with other naval, land and air forces.[1]

France, in particular, will collaborate with Brazil in technology transfer in the design of conventional and nuclear submarines. The perspective of Brazil's Navy is that there should be public private partnership in defense, as the END and the Defense White Paper, encourage the strengthening of relationship of Armed Forces with Brazilian companies so that the production of Brazilian missile meets both national demand and the international market demand.

To cover this historical gap in the Brazilian naval industry, in September 2010, DCNS launched in Cherbourg, the School of Submarines Project, in which engineers and technicians of the Navy of Brazil and the Brazilian companies were selected to absorb the knowledge of submarine project. According to Pierre Quinchon, Director of Submarines Division DCNS, on the opening day of this school, this program confirms the ability of DCNS to conduct innovative partnerships in providing services to its international customers under control technology transfer framework and is a source of pride for the company to allow Brazil to acquire know-how in naval defense domain.

[1] Estratégia Nacional de Defesa. P. 21, http://www.mar.mil.br/diversos/estrategia_defesa_nacional_portugues.pdf

However, Quinchon complements stating that the technology transfer model adopted by the company has the acceptability of all, including the client.[2]

The interpretation of this statement is that the Navy of Brazil, as a customer of DCNS, is aware and is subject to risks associated with the strategic task. The Navy of Brazil, justifying the choice of Scorpène clarified that, some Scorpéne submarine project characteristics deserve special mention. Despite being a conventional submarine, its design is not the evolution of a previous conventional class; on the contrary, the hydrodynamic hull is derived from the nuclear submarine "Rubis / Amethyste" and is more compact. This class of submarine, named Rubis class, has six units in operation in the French Navy.

In addition, it employs technologies used in French nuclear submarines, like the combat system SUBTICS. As a result, the main advantage of our project is to facilitate a rapid transition to nuclear. In interpreting this official explanation it is possible to say that even if DCNS cannot fulfill the agreement and not to transfer the strategic knowhow, the Navy of Brazil can build the submarine with nuclear propulsion themselves as the Navy of Brazil would absorb the knowledge to design and master the nuclear submarine construction technology.

In 2009, the Navy of Brazil created the Coordination General of PROSUB (COGESN) to manage the activities of the program. In this Coordination, under PROSUB, technology transfer is defined as the set of knowledge, information, skills and expertise that provide the customer the required expertise needed to fulfill the Agreement Objective, which is to obtain, by the Brazilian Navy, the first nuclear-powered submarine.

Under the supervision and approval of COGESN, DCNS selects Brazilian companies participating in the technology absorption process. The areas of PROSUB which the Navy aims to nationalize are: safety, air treatment, skill (life support), combat systems, weapons systems, electricity and automation, propulsion system, platform management system, masts, composite materials, hydraulic system, compressed air system and mechanical systems. Many Brazilian companies have expressed interest in participating in the PROSUB.

In September 2009, the General Board of the Navy Material (DGMM) signed all the contractual documents relating to PROSUB, with the Consortium Sepetiba Bay (CBS, made by DCNS and Brazilian construction company Norberto Odebrecht SA (ODEBRECHT), the DCNS,

[2] DCNS a commencé les transferts de technologies vers le Brésil. La Tribune, 16 de setembro de 2010. http://www.latribune.fr/entreprises-finance/industrie/aeronautique-defense/20100916trib000549553/dcns-a-commence-les-transferts-de-technologies-vers-le-bresil.html

ODEBRECHT, and Itaguai naval Constructions SA (ICN), a Special Purpose company (SPC) set up by DCNS, the ODEBRECHT and the Management company of naval Projects (EMGEPRON). This shipping company is the representative of the Navy of Brazil, the Consortium ICN and holds a Special preferred Share called Golden Share[3]. This preferred share is configured as a means of monitoring compliance with the strategic purpose of the SPE.

According a memo was sent by the Navy Command of the Chief of the Navy, the General Commander of the Marine Corps, the Naval Operations Command, the Director General of Marine Material, the Secretary-General of the Navy and the Director- Navy Personnel Overall, stating the relationship between France and Brazil in the signed Cooperation Agreement between the Ministries of Defense of both countries, titled Submarine Development Program (PROSUB) and its Governance, of 8 June 2010.

It covers political, diplomatic and commercial areas[4]. Politically, the relationship is between the Commander of the Navy and Chief of Staff of the Marine Nationale Française (MNF). For the diplomatic level, the relationship is established between the DGMM and the DGA.

According to this Memorandum, the relationship will be developed within the Cooperation Committee of France-Brazil, which is co-chaired by the International Development Director of the DGA, from France, and the Director-General of Naval Material from Brazil. The Committee will have the participation of other ministries (Foreign Affairs, Defense and Finance) of both countries, as set out in Article V of the Cooperation Agreement and detailed in Article 5.3 of the Technical Arrangement. The Committee has, on the French side, a Secretary and the Brazilian side, a Secretariat, coordinated by a Secretary, representative of DGMM, ASSISTED BY TWO Deputy Secretaries, a General Coordination of Submarine Development Program (COGESN) and one of the Navy's Technological Center in São Paulo (CTMSP).[5]

At the commercial level, the relationship between France and Brazil was established by the French part, through Consortia CBS and ICN, and by Brazil, through COGESN. Although COGESN is responsible for implementing the PROSUB, this Coordination is subject to DGMM. In addition to an External Financing Agreement, under the implementation of COGESN, there is a set of seven commercial contracts signed with DCNS, ODEBRECHT, CBS and ICN.

[3]Special share that confers special rights and special statutory provisions and focus on strategic nature of decisions for the company.

[4]Memorandum N° 5/2010 from Navy Command, Programa de Desenvolvimento de Submarinos (PROSUB) e sua Governança, de oito de junho de 2010. Marinha do Brasil.

[5]Memorandum N° 5/2010 from Navy Command Programa de Desenvolvimento de Submarinos (PROSUB) e sua Governança, de oito de junho de 2010. Marinha do Brasil.

In the technology transfer process for conventional submarines, as well as qualification of human resources at various levels and specialties for the construction and detailing of the modified sub section design, technical advice during construction and completion of the project is also included. In addition to professionals who participated in the construction and modernization of submarines, IKL type of Arsenal in Rio de Janeiro Navy (AMRJ); for this activity, the Force also includes business professionals, as Nuclebrás Heavy Equipment S.A. (NUCLEP) and ICN.

The industrial centers of DCNS in France, are: Lorient, where the submarine design is done, Cherbourg, where production is located, Ruelle, where the factory for strategic equipment is located, Nantes-Indret, where propulsion is developed and Toulon, where the submarine combat system is developed. In the technology transfer process, to build the submarine hull with nuclear propulsion, the qualification of engineers of different levels and specialties in France, takes about two years, it was started in July 2012.

As for the technology transfer process for the design and construction of structures for Manufacturing Steel Unit (UFEM) and the Shipyard and Naval Base (EBN), DCNS will be responsible with the Brazilian company ODEBRECHT, to give requirements, technical information, evaluation, certification of design and technical advice during construction.

The scientific and technological facilities of the Navy of Brazil that support the nuclear program of the Navy (PNM), are the pilot plant for the production of UF6 (USEXA) Enrichment Isotope Laboratory (LEI), the Unit Enrichment Pilot Plant (USIDE), the Laboratory of nuclear material (LABMAT), the nuclear-Generation Laboratory (LABGENE), the Storage of nuclear material (ARM) and the Hot cells Pilot Unit (UCQP). These units, operating on the premises of the Experimental Center Aramar (CEA)[6] , since 2012, made it possible for Brazil to master all stages of the nuclear fuel cycle, starting from the extraction of uranium to the conversion and manufacture of the fuel element. The USEXA is the pilot unit which converts the uranium ore to yellowcake and into uranium hexafluoride (UF6) gas; the output of 40 tonnes of natural UF6 per year, is obtained on an industrial scale. LABGENE is responsible for the nuclear reactor for naval propulsion. Its mission is to validate the design conditions and test all possible operating conditions for a nuclear propulsion plant of the PWR type for submarines. This lab will consist of eleven main buildings, including the building for the reactor and the building for the turbines. The reactor model that the Navy of Brazil will use in its nuclear submarines will be the one being developed since the late 1970s; it has three circuits: primary, secondary and the cooling circuit.

[6]Experimental Complex, operated by Centro Tecnológico Marinha de São Paulo

The Brazilian Navy is constructing in the ARAMAR Experimental

Center, located 120 km away from São Paulo, a prototype nuclear installation for generating electricity based on a pressurized water reactor (PWR) of low power [7].

This facility aims in achieving the national technological capability to develop in the country, in the next few decades, nuclear plants of low and medium power and an installation for naval propulsion that satisfy the national defense strategic needs.

The Nucleoelectric Laboratory (LABGENE) was designed with the purpose of validating new concepts and devices that will improve the performance and safety of the nucleoelectric generation and will be a powerful tool for R&D activities relating to fuels and systems that will give support to the development of commercial and naval applications.

An extensive experimental program was carried out for validating the neutronic parameters of the reactor core project..

This program is being implemented in a critical facility specially built for this purpose that started its operation in 1988. This small research reactor, named IPEN/MB-01, was entirely designed by Brazilians and was a remarkable feat of the national engineering at the time.

The main components of the primary circuit, namely the reactor vessel (see figure above) and its internals, pressurizer, cooling pump, steam generators and associated ancillary equipment have been completed. The reactor vessel and its internals have recently been successfully assembled in order to verify adjustments and tolerance.

The main cooling pumps will be subject to performance tests in an experimental facility specially designed for this purpose. This facility is connected to the Experimental Thermo-hydraulic Circuit with pressure of 150 atmospheres (CTE-150), originally designed and previously used for validating the parameters adopted by the project for the primary and secondary circuits of the installation.

The project's parameters and operational performance measurements of the control rods drives mechanisms were also validated in a test bench specially built for this purpose. Other experimental test benches were used for validating the hydro-dynamic parameters and the steam driers. Full scale mock-ups of the reactor and of the equipment for refueling were also constructed for validating the corresponding operational procedures. These experimental test benches and circuits constitute the multi-purpose thermodynamic laboratory (LABTERMO).

The main components of the secondary circuit are also completed, including four turbo-generators, two main condensers that constitute a single unit, four condensate extraction pumps and four water feed pumps.

[7]GUIMARAES, Leonam dos Santos, Completion of Fabrication and Assembly of the Internals and Pressure Vessel of the LABGENE Reactor, http://207.57.4.50/eee53/eee53e/ecen_53e_labgene.htm

The turbo-generators are being tested for their performance according to military rules (MIL-STD) in another laboratory, namely the Laboratory for Testing Propulsion Equipment (LATEP), specially built for that purpose.

Software that simulates the dynamic behavior of the complete installation, including navigation effects, will be in operation at the end of 2015. This simulator will permit the verification of the measurements concerning the integrated performance of the installation (primary and secondary circuits and electric system) and the development of technical specifications and procedures necessary for a safe operation. The CTMSP was successfully completed on July 27 2014.

The successful completion of the design, fabrication and assembly of this structure, together with the fuel elements and the control rod drive mechanisms will constitute a complete PWR type Nuclear Reactor, and will be a unique event in the Southern Hemisphere and it should be a source of great pride for the Brazilian Navy as well as for all Brazilians who believe in the competence of the country regarding the indigenous development of complex technologies.

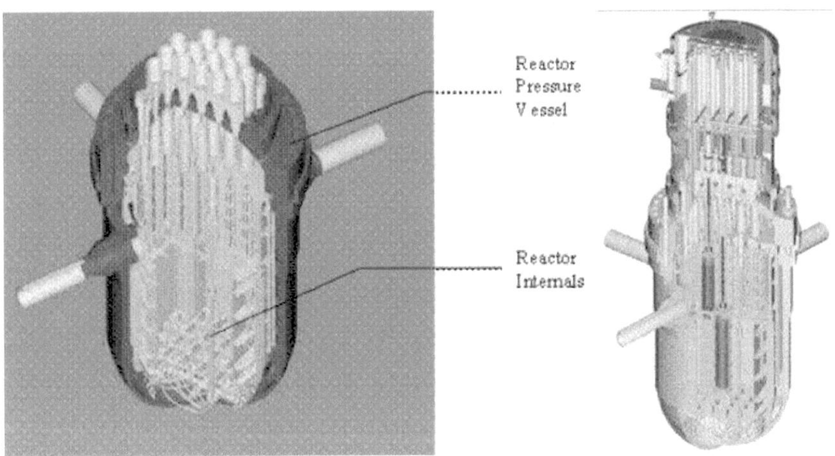

Pressure Vessel and Internals of the LABGENE Reactor

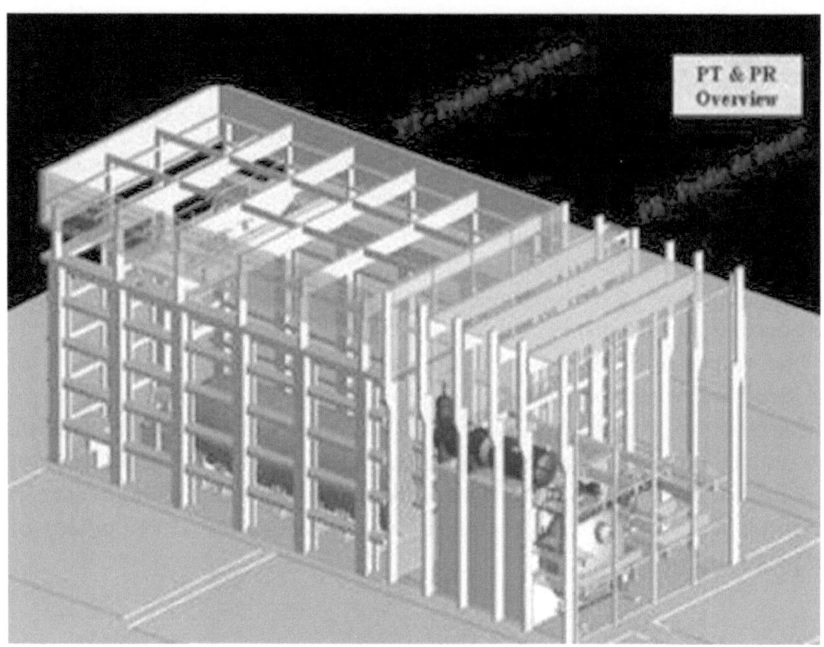

(Top)Reactor and Turbines Building / (Bottom) Reactor vessel with thermal insulation

LABGENE is an undertaking that is being developed under the responsibility of the Brazilian Navy and aims at developing the Brazilian capacity regarding nuclear electric generation which will make it possible to build small nuclear power plants and nuclear naval propulsion systems. The completion of this structure highlights the CTMSP engineering capacity and that of Brazil regarding the design development and practical experience acquired in the fabrication of equipment and this will generate know-how that is unique, innovative and valuable.

From now on the equipment will be stored in the containers and will be kept in an inert atmosphere until the final assembly in the Reactor Building. The development of science and technology, for which creativity and innovation must necessarily be present, is based on three fundamental premises: the first one is the human brain; the second one is to coordinate the persons and institutions around the objectives, aims and targets that generate some strategic or social benefits and the third one refers to the national efforts, allocating adequate resources to the science and technology.

The civil works at the site had been interrupted in 2004. The foundations of the Reactor Building and the Turbine Building (see photos above) are completed. The contracts for continuing these works and for the fabrication and construction of the Nuclear Containment are ready to be signed and they are only awaiting the availability of financial resources. From 1980 until the end of 2004 about US$ 318 million have been invested in LABGENE for the research, development, acquisition and construction and that corresponds to about 65% of its total value. The works were restarted in 2010. It is estimated that US 170 million will be necessary to conclude the LABGENE. Depending on the effective chronogram of allocated resources in the next years, the first reactor criticality could technically be reached in 2017.

One of the main obstacles in the construction of future submarine with nuclear propulsion is the training of human resources. Due to the curtailment of the defense sector, the various budget cuts and the various project activities, the continuation of research in the project was compromised. The solution, in 2012, was to create a new state, the Blue Amazon Defense Technology SA (AMAZUL), linked to the Navy of Brazil and the Ministry of Defense in order to house the human resources allocated to the nuclear program of the Navy (PNM), the PROSUB and the Brazilian Nuclear Program (PNB). This company is the result of the discussions held from the creation of the GNP Development Committee in mid-2008, whose main objective was to set guidelines and goals for this program and oversee its implementation. During a meeting of this committee, the creation of a company called AMAZUL was proposed to solve the problems related to human resources of PNB . The new public company is another state division linked to the MB, called Management Company of Naval Projects (EMGEPRON). The new government started its activities with one thousand and one hundred employees transferred from EMGEPRON. One of AMAZUL goals is to prevent the depletion of strategic human resources due to low salary and the lack of career prospects. Thus, according to Law No. 12,706, of 8 August 2012, the federal government authorized the split of EMGEPRON and transferring its nuclear sector to AMAZUL Until 31 December 2014, there were 1,511 employees in the state, of which 72 employees were moved to the National Institute of Social Security (INSS), considering the requirement of states of São Paulo and Rio de Janeiro. There were a total of 651 company employees in the city of São Paulo, 746 in the city of Iperó and 42 in the state of Rio de Janeiro. Of these about 1500 employees, 543 have college degrees and 154 employees are teachers and doctors. In the view of the Navy, the nuclear industry is renewed as it has also increased the number of Brazilian universities that offer nuclear engineering course.

The services offered by AMAZUL include the development of new technologies, personnel management and knowledge, product marketing, technical services, project management, implementation and management of projects and operation of civilian and naval nuclear facilities. With the AMAZUL facilities and the Brazilian engineers that get trained by Submarines Project School in France, along with the new engineers, will detail the nuclear-powered submarine project.

Conclusion

The Navy of Brazil is of the opinion that the transfer of strategic technology does not take place quickly or according to purely contractual terms; because of this, the strategic partnership with France for the MAN-SUP and the PROSUB will result in acquiring knowledge for several other projects including nuclear and naval areas. Understanding this process of R&D, Brazil will have the advantage and the example of PROSUB will guide the next defense procurement. The national mobilization, in terms of technology, material and human resources that PROSUB have attracted, will ensure the technological development of dual use technology in the country, generating direct and indirect jobs. This will also contribute in maintaining investment in the area of the defense sector, thus strengthening the integrity and national sovereignty and integration of the country into the international system. With the completion of its first submarine with nuclear propulsion, Brazil will join the list of countries to design, build and operate nuclear submarines.

NUCLEAR PROPULSION IN ARGENTINA

Conceiving sensible projects by a developing country in particular if it is of military significance, it has to struggle with its economics, necessities and diplomatic nuances. Argentina is not an exception and its nuclear propelled submarine project has seen several ups and down since its origin.

Argentina possesses reasonable important resources like tight and shale gas, oil, food and water, fishery zones, nuclear technology, lithium resources among many others. Argentine nuclear-propelled submarine project is based on the strategic necessity of warranting its sovereign rights over her huge maritime zone and the natural resources located within the zone. Argentine Sea is a maritime zone of special global interest. This area is of strict application of the rights granted by the Third Convention on Law of the Sea to the coastal country: Argentine Sea is the part of South Atlantic Ocean that covers the continental platform adjacent to the coastal waters of Argentina. According to the Third Convention on Law of the Sea parameters of 1982 and the outcomes of the Fortieth Session of the Commission on Limits of Continental Shelf of March 2016, the Argentine continental platform becomes the most extent underwater flatland of southern hemisphere reaching a total surface of more than 6.5 million square kilometers. This platform progressively increases its width from north to south to more than 2000 kilometers at the 50 S degrees latitude. Without any doubt, the defense of this national asset implies not only strong surveillance and dissuasive naval capacities against third party threats, but also to enforce national sovereignty.

In particular to fish resources, Argentina is in a constant fight with poachers because of this huge extension in its exclusive exploitation area. Modern fishing vessels have similar speeds to military ships and also have warning systems that anticipate the arrival of surface ships, giving them enough time to move to international waters. Thus, the surface fleet is highly inefficient for this task. Similarly the aerial surveillance which has the ability to exercise some police control would be null, or disproportionate. The solution lies in the use of nuclear propelled submarines that could be positioned in the area quietly, even without surfacing any time.

The increasing scarcity of the natural resources required to run an economic activity that integrates a growing world population in a reasonable overall comfort environment is a challenge affecting the international community collectively. In such scenario the potential for conflict is growing; the need for this strategic component of the defense is clear.

Design Philosophy

The general conception of the components of defense has to do with a military policy based on the role that a country intends to play. Argentina does not intend to achieve any strategic military advantage. On the contrary, Argentina assumes a non-provocative defense attitude. The spirit of a non-provocative defense military policy lies behind a military power that allows an effective defense of the nation, but by no means having the opportunity to initiate or sustain full scale military operations over other countries territories or assets. But, nonetheless, it conceives a moderate counteroffensive capacity as a dissuasive component.

Argentine nuclear propelled submarine concept arises from above. A single nuclear-propelled attack submarine with conventional weapons does not fulfill the minimum conditions to become a full-scale warfare threat. It will be limited to an access-denial doctrine over the South Atlantic Ocean and a fully geographical naval retaliation capacity as a key dissuasive component.

Argentina fosters initiatives that promote local research to achieve full mastering of its own nuclear component for the submarine. That has been a must from the beginning of the project. Besides, Argentina is going to upgrade their actual TR-1700 class submarine platform.

Argentine nuclear propulsion submarine is expected to have the following characteristics:

Type: Nuclear-powered interception and attack
Platform: TR-1700-class diesel-electric
Displacement: 3.000 tons, submerged
Operating depth: 300 meters
Range: unlimited
Operating period: 90 days
Length: 66 meters
Beam: 9 meters
Draught: 8 meters
Fuel: enriched uranium 20%
Propulsion: Nuclear-electric, CAREM-based nuclear reactor ~ 2 MWe, 150 KW LNG aux electricity generator, 1200 battery elements, Electric motor ~ 7 MWe, 1 screw
Speed: 26 knots submerged
Armament: 6 torpedo tubes, up to 24 SUT-264 or SEAHAKE MOD 4 torpedoes.

History Of The Project

Towards the end of 1974, the Argentine government signed an agreement with the German company Ingenieur Kontor Lübeck (IKL), now a part of the Thyssen Nordeseewerke industrial conglomerate, to design and build a new family of submarines in order to replace Argentine

Submarine Force (COFS) Guppy and U-209 boats. Argentine Navy's specifications were tailor-made; a truly ocean attack submarine that could be upgraded to nuclear propulsion in the future. The outcome was TR-1700/1400-class concept submarine, providing outstanding hydrodynamic performance, 300 meters submerged operating depth, sustained submerged speed of 25 knots for more than an hour; 22.000 kilometers of autonomy and very low underwater noise signature. Such features were far superior to those of any conventional submarine that were available at that time.

Finally in April 1977, Argentine government signed off contracts with both Thyssen Rheinstahl Technik and Thyssen Nordseewerke to provide services and supplies required for the installation of a shipyard to be located in Argentina named after Domecq García Shipyard), provision of one TR-1700-class submarine built in Germany, services and supplies to locally build three TR-1700-class submarine and the same for two TR-1400-class units, and with AEG-Telefunken to provide SST-4 torpedoes for a total amount of D$M 950 million (Deutsche Mark 1977 value).

Eventually the two first class TR-1700 boats were built in Germany: ARA Santa Cruz (S-41) that entered service towards December 1984 and ARA San Juan that entered service towards January 1986. After Malvinas/Falklands' War Great Britain tried to stop the transfer due to the latest conflict.

Construction work of the local own shipyard began during 1978 and officially started its activities in January 1982. As of 1983 the Argentine government entrusted INVAP national agency to carry out a technical feasibility study to design, build and accommodate a nuclear reactor in the submarine ARA Santa Fe (S-43) which was getting constructed. The technical exercise was successful; the proposal was to increase the hull length by 7 meters to make it 73 meters instead of the original 66 meters of the diesel version. This configuration was very similar to French RUBIS class submarines.

The economic and political reasons made it impossible to finish the submarines under construction, even though one of these ARA-Santa Fe (S-43) submarine was 70% completed; the other, ARA Santiago del Estero (S-44) was 50% completed, while the last boats S-45 and S-46 were at very initial stage of construction.

In the 1990s the government of President Carlos Menem announced the closure of not only the nuclear propelled submarine project but the whole Argentine submarine program. Even worse, for some unfathomable reason, there was an idea to convert the shipyard into a shopping mall. The shipyard was scrapped; the tools and hardware was transferred in part to the Navy while the remaining equipment was sold at a throwaway price. The submarines under construction were abandoned for more than 20 years. The government of President Carlos Menem was responsible for the shipwreck of those strategic defense-related industries, among many others. Rumor was that it was part of Great Britain's non-official imposed conditions for renewing diplomatic relations after Malvinas/Falklands' War. These relations were not crucial at all for Argentina, but were a personal prestige for the president to show a diplomatic achievement similar to that of his predecessor Raúl Alfonsín on the conflict with Chile over the Beagle Channel.

Most recently in 2004, the Argentine government of Néstor Kirchner reopened the shipyard, renaming it as Almirante Segundo Storni Shipyard. At the same time, TANDANOR naval workshop was brought back under State control. Both were put together as a new naval conglomerate called CINAR (Argentine Naval Industrial Complex) with its main objective and mission of re-establishing naval, engineering and industrial resources and capacities. First submarine related activity was the maintenance works on U-209 boats ARA Salta (S21) and ARA San Luis (S-22). Later challenge was to do the mid-life maintenance work on ARA San Juan (S-42) TR-1700-class boat needing modernization.

By mid-2010, Argentine Minister of Defense Mrs Nilda Garré announced a new nuclear propulsion program to power Argentine Navy ships; everyone assumed that the first nuclear-powered unit would be a submarine. Once again INVAP agency was endeavored to build a new proposal on this issue. The nuclear plant would be based on the ongoing CAREM (Argentine Central of Modular Elements) reactor project (SMR, Small & Medium Reactor project).

This time it was considered a nuclear-electric configuration which basically means to replace the diesel generators by a nuclear generator for charging the batteries. Under this configuration there would be no need to expand extend submarine's hull is to be needed.

Finally, in 2011 it was proposed that the work on the rusty ARA Santa Fe (S-43) would be resumed; to be fitted with nuclear propulsion.

AR's Carem Reactor

Argentine Nuclear Program considers developing a nuclear power reactor called CAREM (Argentine Central of Modular Elements) amongst its projects. This would bring the possibility for a further development of a compact nuclear power reactor to power the submarine.

The key goal of CAREM project is to completely design, build and set up a power nuclear reactor prototype with at least 70% supplies and related services provided by Argentinian companies.

This type of reactors has a huge future projection for electricity supply over rural and remote areas far away from urban zones and for high-consuming industrial conglomerates. Main current benefits concern water desalination, vapor production for industrial purposes.

CAREM design has two crucial aspects that simplify its building, setting up, operation and maintenance. The first, its passive safety systems, the second, its total-integrated format with reactor core, control rod drive, steam generators all internal accommodated in a single reactor pressure vessel.

An interesting feature of the CAREM reactor is that it does not include core coolant pumps.

CAREM coolant is light water, which also acts as moderator.

Inside the pressurized vessel, water flow is entirely by "natural convection". Natural circulation of water is caused due to the temperature gradient inside the vessel and its sources located at different heights. The reactor cores are called "hot source" is located at the bottom of the vessel while the "cold source", the steam generators, are at the top.

Another advantage of the integration is the reduction of also the size of components as well as the risk of incidents.

The CAREM nuclear reactor will include four main components: reactor core, primary circuit, pressure vessel and secondary circuit.

The core has 61 fuel elements entirely engineered by CNEA (Atomic Energy National Agency) and already tested in the RA6 research reactor (another entirely local development). Each of the fuel elements includes more than a hundred enriched (3.5%) uranium rods plus radiation-absorbent materials to keep the nuclear fission process under control.

In case of the compact future version for the submarine, fuel should be enriched up to 20%, maximum considered by NPT as non-proliferation use of nuclear energy. Besides, submarine's fuel would necessarily have its lifetime equivalent to the submarines and avoid the very complex process of refueling.

The primary circuit of the CAREM makes possible the coolant circulation possible, which both cools and transfers heat to steam generators. Coolant is light water.

The pressure vessel is designed to operate in high temperatures and high pressure conditions. Its 20 cm thickness wall is made of forged steel and has Stainless steel clad on the internal surface. The vessel is located in a thick structure of concrete and steel which can collect any leakage and also acts as shielding

The secondary circuit includes 12 steam generators, a steam turbine and a coolant circuit for water circulation.

The Safety of the CAREM reactor will be assured by both passive systems, which reduce risk of incidents, and active systems, installed as a backup. Safety systems have the function of ensuring the stoppage of the fission process, removes residual heat from the core and keeps the reactor off. Safety systems also comply with requirements of redundancy, independence, placement, diversification and self-operation.

The reactor's core includes a "fast acting shutdown" system of additional control rods that will always be fully withdrawn during operation, but which will scram when required for shutdown together with rods of the control system thus stopping the fission reactions in a few seconds and keeping the reactor shutdown. In case of an emergency shutdown the control and shutdown systems are backed up by borated water injection systems, passive residual heat removal systems and external safety water injection systems.

Agencies Involved, Industrial Resources

Argentina is the most advanced country in Latin America on nuclear technology. Since 1950 it is pursuing the study, research and application of nuclear technology on a wide range of peaceful applications, in spite of the numerous difficulties and conflicts during the last decades. However, Argentina strongly supports the use of nuclear technology for peaceful purposes, with utmost attention on energy generation, medical applications and nuclear fuel cycle from mining to waste disposal. The use of nuclear technology for propulsion is considered a peaceful application; a propulsion system does not release uncontrolled and large quantity of energy like a nuclear weapon nor it result in any nuclear contamination.

Nuclear-related activities in Argentina are centralized in CNEA (National Atomic Energy Agency) founded in 1950. It is an autonomous agency reporting to the Ministry of Energy and Mining; the powers and functions are defined by National Nuclear Activity Law number 24804 and amendments thereto. During its six decades, CNEA carried out nuclear related activities with different agencies and companies; among these are DIOXITEK S.A. to manufacture uranium dioxide and cobalt-60 sealed sources, PIAP for heavy-water production, CO.NU.AR S.A. for manufacturing fuel elements and control rods used in the three Argentina nuclear plants Atucha I, Atucha II and Embalse Río III, FAE S.A. and manufacturing zirconium-alloy tubes and rods (used by CO.NU.AR for fuel elements) as well as stainless steel and titanium-alloy seamless pipes. N.A.S.A. (Nucleoeléctrica Argentina S.A.) is in charge of the building, operation, maintenance and decommission of Argentine nuclear plants, PILCANIYEU Technological Center is dedicated to uranium enrichment through gaseous diffusion method.

INVAP S.E. was founded to build complex engineering systems. By now it has achieved a vast experience in building nuclear research reactors and radioisotope production nuclear reactors, and has also exported to Peru, Australia, Egypt and Algeria. INVAP is also engaged in radar and satellite research and manufacturing.

CNEA closely works with other national agencies too, like INTI (National Industrial Technology Agency), INTA (National Agricultural

Technology Agency), CITEDEF (Defense Scientific and Technological Research Agency), SEGEMAR (Argentine Geologic Mining Service) and UNSAM (National University of San Martin).

ARN (Nuclear Regulatory Authority) is in charge of establishing and enforcing the regulatory framework regarding all the nuclear activities in Argentina. Its mission is to protect people, the environment and future generations from the health hazards associated with ionizing radiation. So, the regulatory framework for the use of a nuclear reactor in a submarine as its propulsion system will be of its supervision.

The naval industrial resources for submarine production would reside with CINAR (Argentine Naval Industrial Complex) located in Buenos Aires City. CINAR was founded in 2010 uniting TANDANOR Naval Workshop with ALTE STORNI Shipyard (formerly DOMECQ GARCIA Shipyard) as a unique naval-industrial complex. The history of TANDANOR Naval Workshop dates back over 130 years to 1879 when it was set up as the Naval Workshop for the Argentine Navy's ships maintenance and repair.

The future Argentine nuclear propelled submarines are expected to be commissioned to the Argentine Navy's Submarine Force Command (COFS) which is actually land-based in Mar del Plata Naval Base (BNMP) located in Mar del Plata City in Buenos Aires province. However, this naval base is not suitable to support nuclear propelled submarine operations due to the issues of nuclear security and safeguard. To solve this, COFS is planned to be transferred to Caleta Paula Naval Base (BNCP) situated near Caleta Olivia City in Santa Cruz province in the south of Argentina. The port of Caleta Paula also offers modern and secure facilities for oil tankers, bulk carries or fishing vessels maneuvering; and it is the continental middle point of Argentine maritime jurisdiction over the South Atlantic Ocean.

Commitment to the Peaceful Use of Nuclear Energy

Joining the "nuclear weapons club" was never been an aspiration of Argentina. However, a strong position with respect to nation's autonomy in nuclear energy development for peaceful uses is supported by Argentina. It was so strong that it was misunderstood as an intention to obtain a bomb during the 1970s.

Most of the countries signed the NPT, thus committing for the international controls on nuclear issues. Argentina, because of the mistrust of the Brazilian nuclear programme and vice versa, had necessarily to undergo a different position with its neighbor to defuse the political situation. In the beginning of 1980s, negotiations between the two countries were started seeking a common bilateral strategy in this central area of international security. In 1985 the both democracies of Argentina and Brazil issued the Joint Statement on Nuclear Policy. Bilateral talks ended in July 1991 signing the Guadalajara Agreement, declaring exclusive use of nuclear energy for peaceful purposes and the establishment of the Brazilian-Argentine Agency for Accounting and Control of Nuclear Materials (ABACC) clearly demonstrating the political will of both countries to give transparency to their nuclear programs contributing to build mutual trust. Immediately after, in December 1991 the Quadrilateral Agreement among Brazil, Argentina, ABACC and the International Atomic Energy Agency (IAEA) committed both countries to adhere to international safeguards over all nuclear materials and activities. This scheme is similar to the one, used among IAEA, European Atomic Energy Community (EURATOM). and its non-nuclear weapons members. Finally, Argentina adhered to the NPT in 1995 and later in 1998 both countries ratified the Convention on the Physical Protection of Nuclear Material (CPPNM) of 1980 and the Comprehensive Nuclear Test-Ban Treaty (CTBT) of 1997 and also adhered and ratified the 2005 Amendment to the Convention on the Physical Protection of Nuclear Material. Thus, Argentina has demonstrated to the international community its intentions of peaceful use of this technology, though, being the country that only uses natural and slightly enriched uranium in its programme; Argentina has proved itself trustworthy and predictable on nuclear issues, rigorously observing international norms concerning non-proliferation.

Project's Timeline

The first attempt was supposed to deliver the first nuclear-propelled submarine by mid 1990s; it was halted in the early 1990s. A decade later, in 2010, Ministry of Defense announced reopening the project.

Feasibility studies conducted at both times clearly indicated that the Argentine nuclear-propulsion submarine was technologically within its reach; of course, the political 'will' is necessary, in addition to the budget.

Argentina needs a thorough reform and restructuring of its military apparatus as well as a profound social discussion on Defense matters and the role of its Armed Forces in the progress of the society, especially in the political sphere that mainly holds a supine ignorance on this issue. Meanwhile, Argentine nuclear propelled submarine will remain an unsatisfied strategic national need.

Special thanks to: VGM VAlte (RE) Antonio Mozzarelli, Prof. Thomas Scheetz.

CIVILIAN SUBMARINES

Although the majority of the world's submarines are military, there are some civilian submarines, which are used for tourism, exploration, oil and gas platform inspections, and pipeline surveys. Some are also used in illegal activities.

The Submarine Voyage ride opened at Disneyland in 1959, but although it ran under water it was not a true submarine, as it ran on tracks and was open to the atmosphere

The Mésoscaphe Auguste Piccard, built by Jacques Piccard for the 1964 Swiss national exhibition, is the first tourism submarine in history. It transported some 33,000 tourists through the depths of Lake Geneva during 1964–65. By 1997 there were 45 tourist submarines operating around the world. Submarines with a crush depth in the range of 400–500 feet (120–150 m) are operated in several areas worldwide, typically with bottom depths around 100 to 120 feet (30 to 37 m), with a carrying capacity of 50 to 100 passengers.

In atypical operation a surface vessel carries passengers to an offshore operating area and loads them into the submarine. The submarine then visits underwater points of interest such as natural or artificial reef structures. To surface safely without danger of collision the location of the submarine is marked with an air release and movement to the surface is coordinated by an observer in a support craft. A recent development is the deployment of so-called narco submarines by South American drug smugglers to evade law enforcement detection. Although they occasionally deploy true submarines, most are self-propelled semi-submersibles, where a portion of the craft remains above water at all times. In September 2011, Colombian authorities seized a 16-meter-long submersible that could hold a crew of 5, costing about $2 million. The vessel belonged to FARC rebels and had the capacity to carry at least 7 tonnes of drugs.

REFERENCES AND BIBLIOGRAPHY

Front Cover image: USS Dallas departs Souda Bay, U.S. Navy photo by Paul Farley.

Wikipedia: Many chapters and images of this book extensively use Wikipedia as the source of information. It is not possible to use all the links from Wikipedia.

Others:

- Submarine History Year 1580-2000
- Nuclear Navy by Hewlett
- Atomic submarine by Blair
- Nuclear Propulsion for merchant ships by Krammer
- Concept in submarine design-Burcher, Roy
- Submarine geology by Shepard , Francis P
- On the crisis stability of a submarine Deterrent by Gregory H. Canavan
- Navy Ohio Replacement (SSBN[X]) Ballistic Missile Submarine Program: Background and Issues for Congress Ronald O'Rourke
- Yogesh Joshi-PhD thesis—From Denial to Deterrence ; India's Quest for a Nuclear Submarine –School of international Studies, Jawahar Lal Nehru University
- Joseph P. Chacko-Foxtrot to Arihant
- India's Nuclear Submarine Programme Rear Adm AP Revi
- Anil Anand-The Second Strike
- K Raghuraman, former Science councilor in the Indian Embassy in Vienna—Presentation to the Trombay Council, BARC.
- Fraunhofer Institute of Optronics, System Technologies and Image Exploitation IOSB

From the net

- Submarine reactor compartments
- Arihant
- Arihant Class submarines-Wikipedia
- Arihant=ATV -Global Security.org

Brazil Chapter

- CORRÊA, Fernanda das Graças Corrêa. O projeto do submarino nuclear brasileiro. Uma história de ciência, tecnologia e soberania. Rio de Janeiro: Capax Dei, 2010.
- GUIMARÃES, Leonam dos Santos. Estratégias de Implementação e Efeitos de Arraste dos Grandes Programas de Desenvolvimento Tecnológico nacionais: estudo de caso do programa de propulsão nuclear da Marinha. Cadernos. Faculdades Integradas São Camilo, São Paulo, v. 10, n.3, 2004.
- GUIMARAES, Leonam dos Santos, Completion of Fabrication and Assembly of theInternals and PressureVessel of the LABGENE Reactor, http://207.57.4.50/eee53/eee53e/ecen_53e_labgene.htm
- Memorando Nº 5/2010 do Comando da Marinha Programa de Desenvolvimento de Submarinos (PROSUB) e sua Governança, de oito de junho de 2010. Marinha do Brasil.
- OLIVEIRA, Eliézer Rizzo de. Democracia e defesa nacional. Barueri, São Paulo: Manole, 2005.
- Acordo entre o Governo da República Federativa do Brasil e o Governo da República Francesa na área de submarinos. 23 de dezembro de 2008. Disponível:
- http://npsglobal.org/esp/images/stories/pdf/Acuerdo_Brasil_Francia.pdf
- Apresentação de Nelson Jobim para o Conselho de Administração da Odebrecht. São Paulo, 24 de maio de 2011. https://www.defesa.gov.br/arquivos/File/2011/mes05/futuro_defesa1.pdf

- CTMSP. https://www1.mar.mil.br/ctmsp/labgene
- DCNS a commencé les transferts de technologies vers le Brésil. La Tribune, 16 de setembro de 2010.
- http://www.latribune.fr/entreprises-finance/industrie/aeronautique-defense/20100916trib000549553/dcns-a-commence-les-transferts-de-technologies-vers-le-bresil.html
- Entrevista com o Almirante Hecht, gerente do PROSUB, em 15 de fevereiro de 2011. http://www.slideshare.net/telmamonteiro/submarinos-1979235
- Estratégia Nacional de Defesa. Disponível:
- http://www.mar.mil.br/diversos/estrategia_defesa_nacional_portugues.pdf
- Jobim defende transferência de tecnologia para militares. Agência Câmara de Notícias, 31 de outubro de 2007.
- http://www2.camara.leg.br/camaranoticias/noticias/112889.html
- Lorient: Une école de conception pour les sous-marins brésiliens. Mer et Marine, 17 de setembro de 2010. http://www.meretmarine.com/article.cfm?id=113975&u=77291
- Lula e Sarkozy assinam acordos da área militar. Notícias JusBrasil, 23 de dezembro de 2008. http://www.jusbrasil.com.br/politica/633008/lula-e-sarkozy-assinam-acordos-da-area-militar
- Presidente francês parabeniza Lula da Silva e elogias conquistas sociais. BBC Brasil, 30 de outubro de 2006. http://www.bbc.co.uk/portuguese/noticias/story/2006/10/061030_lulachiracaw.shtml
- Sarkozy anuncia volta da França ao comando militar da Otan. Estado de São Paulo, 11 de março de 2009.
- http://www.estadao.com.br/noticias/internacional,sarkozy-anuncia-volta-da-franca-ao-comando-militar-da-otan,337135,0.htm
- Scorpene DCNS.A new benchmark for performance.
- http://en.dcnsgroup.com/naval/products/scorpene/?product-category=ssks
- Scorpène: Fin du litige entre DCNS et Navantia. Mer et Marine, 15 de novembro de 2010.
- http://www.meretmarine.com/article.cfm?id=114506&u=77291
- Submarino Scorpène: a posição da Marinha - Submarinos na estratégia naval brasileira. Defesanet, 22 de dezembro de 2008.
- http://www.defesanet.com.br/prosub/noticia/1926/SUBMARINO-SCORPENE--A-POSICAO-DA-MARINHA-Submarinos-na-estrategia-naval-brasileira
- ou http://www.mar.mil.br/menu_v/ccsm/temas_relevantes/submarino_Escorpene.html

France Chapter

ARCEA GASN
La Propulsion Nucléaire
Bulletin n°43 - 8 May 2010
http://fr.calameo.com/read/0004132413b79a7dbab61

Commissariat à l'énergie atomique et aux énergies alternatives (CEA)
Histoire du CEA – Les Dates Clés
Juillet 2012
http://www-cadarache.cea.fr/fr/accueil/pdf/CEA_livret_Histoire_1945-2012.pdf

CEA
A terre comme en mer
ATOUT Cadarache - n°3 - Mars/Avril 2004
http://www-cadarache.cea.fr/fr/activites/pdf/Dossier_ATOUT_EXT_n3.pdf

CEA
Histoire du Centre de Saclay
Website of the CEA Saclay Nuclear Research Centre
http://www-centre-saclay.cea.fr/fr/Histoire-du-centre-de-Saclay

DCN
Les Redoutable, histoire d'une aventure technique, humaine et stratégique
Website of the magazine Mer et Marine
http://www.meretmarine.com/fr/content/les-redoutable-histoire-dune-aventure-technique-humaine-et-strategique-0

FRIBOURG Charles
La Technologie des Réacteurs de Propulsion Navale
from IAEA – INIS base – Non referenced
http://www.iaea.org/inis/collection/NCLCollectionStore/_Public/33/048/33048066.pdf

GEMPP André
La Mise en Place et le Développement des Sous-Marins Nucléaires
Institut de Stratégie Comparée, Paris
http://www.institut-strategie.fr/ihcc_nuc1_Gempp.html#Note1

GOLDSCHMIDT Bertrand
L'aventure atomique
Fayard, 1962

LEFEBVRE Véronique
Au coeur de la matière. 50 ans de recherches au CEA de Saclay
Cherche Midi, 2002

LOVERINI Marie-Josée
L'atome. De la recherche à l'industrie, le commissariat à l'énergie atomique
Découvertes Gallimard, 1996

Marine Nationale
Programme Barracuda : Mer en vue pour le Suffren
Website of the French Navy – May 2013
http://www.defense.gouv.fr/marine/magazine/traque-en-eaux-profondes/programme-barracuda-mer-en-vue-pour-le-suffren

Ministère de la Défense DGA – DCN
La génèse de la propulsion nucléaire en France
in Proceedings of the Conference*"1899 / 1999, un siècle de construction sous-marine"*
Cherbourg, 25-26 October 1999
http://www.sous-mama.org/la-genese-de-la-propulsion-nucleaire-en-france-blog-254.html

MOULIN Jean, ISNARD Jacques
De la mer à la terre : Les enjeux de la Marine française au XXIe siècle
Editions Perrin, 2006

Pinault Michel
Frédéric Joliot-Curie
Odile Jacob, 2000

PO Jean-Damien
Les moyens de la puissance - Activités militaires du CEA (1945-2000)
Ellipses – Perspectives Stratégiques, 2001

Ville de Fontenay-aux-Roses
Et, Zoé est arrivée
Bulletin Liens de mémoire n° 11 – Pages d'Histoire 2nd semester 2008
www.fontenay-aux-roses.fr/fileadmin/...la.../Histoire_du_CEA.pdf

Wikipedia
Histoire du programme nucléaire militaire de la France
https://fr.wikipedia.org/wiki/Histoire_du_programme_nucléaire_militaire_de_la_France

ABOUT THE AUTHOR

Anil Anand did his mechanical engineering from BHU in 1961 and joined Atomic Energy, where he did his post-graduation in Nuclear Science and Technology.

At the time of retirement from Dept. of Atomic Energy on 30.11.2001, he was Director, Technical Coordination and International Relations Group, Bhabha Atomic Research Centre, and DIRECTOR, Reactor Projects Group, leading the Programme for Nuclear Propulsion for the Indian Nuclear Submarine Programme. He was also the Executive Director, Kudankulam Project, in Nuclear Power Corporation.

After retirement from BARC, he was Scientific Consultant, in the Office of the Principal Scientific Advisor to the Government from February 2002 to February 2008

At present, he is Director Technical, Microtrol Sterilisation Services Private Ltd. Mumbai where he designs the Ethylene Oxide, Gamma sterilization plants for, sterilization of disposable medical devices and for bio burden reduction of Spices.

His Biography/Autobiography 'THE SECOND STRIKE' was published and released by Director BARC on September 14, 2014. This book is available as 'Ma Dame - A Nuclear Scientist's Tryst with Love and Fission' (ISBN-13: 9789385699023) for international markets.

Some comments on the book:

Thank you very much for remembering me and sending me your just published 'Memoir' ! It brought back so many happy memories of our meetings and interactions .You have articulated a number of important issues in a subtle manner and I was happy that you stood by your position in the latter years too. We do have to go ahead and your writing will no doubt be an inspiration to the younger generation. - **Admiral Vishnu Bhagwat, former CNS**

just finished reading your book. Honestly, I did not expect it to be so engrossing-it being an autobiography. But your unique and sincere love story, with a foreign element added, grows on one's mind. The transition from Lahore to Bombay, and the struggles through your academic rise, are also inspiring. Though, i could not relate very well with some of your professional technicalities, various departments and their acronyms, i greatly admired your guts and wondered how one person can have so many achievements to his credit. The winding up of the book is very sweet and sober, with love maturing and intensity growing more and ever more. Priceless!

I am glad to have read this book and feel that i know you better now. Congratulations! - **Vijay Sadanah, film & TV producer & director**

I have already read the book and enjoyed every bit of it, especially the story of your remarkable love for your wife, and hers for you, told with so much humour, sincerity and time-enriched poignancy. The tales of your large family and many friends hold their own interest. As for the nuclear bits, I confess I had to read between the lines at times. You remain as guarded as ever to talk about your own great contribution to India's nuclear progress in those trying decades. - **Kiran Doshi, Former Indian Ambassador to Vienna; a writer 'Birds of Passage', a satire on Indo-Pak relations and 'Jinnah often came to our house', a fiction with Jinnah's reality.**

I have started reading the Book and have reached your era at BHU. The narration is excellent. I liked your narration of the developing romance while in France with your remarkable partner. We cannot forget our 2 visits to Bangkok when we stayed in your wife's parental old house and then in her apartment where her sister also had an apartment. Is she still in the same place? I will read every word of this fascinating Book. – **Dr. M S Chadha, former Director Biosciences BARC.**

I've just started to read your book. It's very engaging! - **Ms Sandhya Sharma. HRD Consultant and Guest Faculty at Greenwich University.**

I got halfway through the book and then got very busy, so am carrying it in my workbag and reading it in bits. It is a lovely book - **Jane Bhandari, a writer and a poet.**

Thank you for sharing this BIG narrative.

Heartiest Congratulations on these milestone accomplishments. It must feel tremendously satisfying- professionally and personally.

We share our pride with dear Meera and our families.

We hope the Government of India will honour you with an appropriate and deserving civilian honour. - **Indu and Vishwa Mitter Bhargava from USA**

ABOUT THE CONTRIBUTORS

Argentina :

Alejandro Delaygue is Bachelor in Electronics Engineering with master degrees in Business Administration and National Defense. Worked as Productive Investments Analyst and Commerce in government dependencies, and as Information Technologies consultant, management and productive processes, and general management of medium & small companies concerning services and agro production. Consultant and researcher on defense issues and national strategic industrial development.

Brazil:

Fernanda das Graças Corrêa is the technical advisor at state company AMAZUL Defense Technologies. PhD in Political Science by Fluminense Federal University and in Policy and Strategy from the National War College. Master in Comparative History at Federal University of Rio de Janeiro. Specialist in Brazilian Military History by Federal University of the State of Rio de Janeiro and graduated in History from the University Gama Filho.

Leonam dos Santos Guimaraes - After completing a BSc in Naval Sciences from the Brazilian Naval Academy in 1980, followed by sea duty appointments, he went on to study Ocean Engineering at University of São Paulo and earned a MSc degree. Following a nuclear engineering position at INSTN (University of Paris XI) and a politics and strategy degree at Brazilian Naval War College, Dr. Guimaraes earned his PhD on Naval Nuclear Propulsion Safety from University of São Paulo in 1999. Currently he is Director of Planning, Management and Environment of ELETROBRÁS ELETRONUCLEAR. He was Technical and Commercial Director of AMAZUL Defense Technologies, Coordinator of the Brazilian Navy Nuclear Propulsion Program, and Professor of Operations Management and Risk Engineering at University of São Paulo. He is member of WNA Board, member of IAEA Standing Advisory Group on Nuclear Energy (SAGNE) and President of Latin American Section of American Nuclear Society.

France

Yves M. HÉNON, After his 'Master of Science' from Louisiana State University in the US, Mr. Henon joined the French Atomic Energy Commission at the Cadarache research center in 1980. After working with this governmental organization for seven years, he moved to the private industry, specializing in building and managing facilities using radioisotopes in France, Thailand and Malaysia. He established his own consultancy firm specialized in radiation processing in 2006 in Singapore. After a two-year assignment at the Department of Nuclear Applications at the IAEA, he is the principal of YH Consulting in France where he has now settled. For several years he has been actively involved with the activities of the International Irradiation Association.

www.ingramcontent.com/pod-product-compliance
Ingram Content Group UK Ltd.
Pitfield, Milton Keynes, MK11 3LW, UK
UKHW041828200726
13854UKWH00002BA/884

9 798201 310172